NATURAL DISASTER RESEARCH, PREDICTION AND MITIGATION

FEMA'S DISASTER DECLARATIONS

FACTORS, COST CONSIDERATIONS AND TRENDS

NATURAL DISASTER RESEARCH, PREDICTION AND MITIGATION

Additional books in this series can be found on Nova's website under the Series tab.

Additional E-books in this series can be found on Nova's website under the E-book tab.

GOVERNMENT PROCEDURES AND OPERATIONS

Additional books in this series can be found on Nova's website under the Series tab.

Additional E-books in this series can be found on Nova's website under the E-book tab.

NATURAL DISASTER RESEARCH, PREDICTION AND MITIGATION

FEMA'S DISASTER DECLARATIONS

FACTORS, COST CONSIDERATIONS AND TRENDS

MICHAEL BECK
AND
CARL ROBBINS
EDITORS

nova publishers
New York

For permission to use material from this book please contact us:
Telephone 631-231-7269; Fax 631-231-8175
Web Site: http://www.novapublishers.com

Additional color graphics may be available in the e-book version of this book.

Library of Congress Cataloging-in-Publication Data

ISBN: 978-1-62808-188-6

Published by Nova Science Publishers, Inc. † New York

CONTENTS

PREFACE

The growing number of disaster declarations (a record 98 in fiscal year 2011 compared with 65 in 2004) has contributed to increased federal disaster costs. FEMA leads federal efforts to respond to and recover from disasters and makes recommendations to the President, who decides whether to declare a disaster and increase the usual federal cost share of 75 percent. This book addresses the number of declarations requested and approved from fiscal years 2004-2011 and associated Disaster Relief Fund obligations; the criteria FEMA uses to recommend a declaration for public assistance, and the extent that FEMA assesses whether an effective response to a disaster was beyond the capabilities of state and local governments; how FEMA determines whether to recommend cost share adjustments, and their costs; and FEMA's administrative cost percentages for declarations.

Chapter 1 – The Robert T. Stafford Disaster Relief and Emergency Assistance Act (referred to as the Stafford Act - 42 U.S.C. 5721 et seq.) authorizes the President to issue "major disaster" or "emergency" declarations before or after catastrophes occur. Emergency declarations trigger aid that protects property, public health, and safety and lessens or averts the threat of an incident becoming a catastrophic event. A major disaster declaration, issued after catastrophes occur, constitutes broader authority for federal agencies to provide supplemental assistance to help state and local governments, families and individuals, and certain nonprofit organizations recover from the incident. The end result of a presidential disaster declaration is well known, if not entirely understood. Various forms of assistance are provided, including aid to families and individuals for uninsured needs and assistance to state and local governments and certain non-profits in rebuilding or replacing damaged infrastructure. The amount of assistance provided through presidential disaster declarations has exceeded $140 billion. Often, in recent years, Congress has enacted supplemental appropriations legislation to cover unanticipated costs. While the amounts spent by the federal government on different programs may be reported, and the progress of the recovery can be observed, much less is known about the process that initiates all of this activity. Yet, it is a process that has resulted in an average of more than one disaster declaration a week over the last decade. The disaster declaration procedure is foremost a process that preserves the discretion of the governor to request assistance and the President to decide to grant, or not to grant, supplemental help. The process employs some measurable criteria in two broad areas: Individual Assistance that aids families and individuals and Public Assistance that is mainly for repairs to infrastructure. The criteria, however, also considers many other factors, in each category of assistance, that help decision makers assess the impact of an event on communities and states. Under current law, the decision to issue a declaration rests solely with the President. Congress has no formal role, but has taken actions to adjust the terms of the process. For example, P.L. 109-295 established an advocate to help small states with the declaration process. More recently, Congress introduced legislation, H.R. 3377, that would direct FEMA to update some of its criteria for considering Individual Assistance declarations. Congress continues to examine the process and has received some recommendations for improvements. Given the importance of the decision, and the size of the overall spending involved, hearings have been held to review the declaration process so as to ensure fairness and equity in the process and its results.

Chapter 2 – The growing number of disaster declarations—a record 98 in fiscal year 2011 compared with 65 in 2004—has contributed to increased federal disaster costs. FEMA leads federal efforts to respond to and recover from disasters and makes recommendations to the President, who decides whether to declare a disaster and increase the usual federal cost share of 75 percent. This report addresses (1) the number of declarations requested and approved from fiscal years 2004-2011 and associated DRF obligations; (2) the criteria FEMA used to recommend a declaration for PA, and the extent that FEMA assessed whether an effective response to a disaster was beyond the capabilities of state and local governments; (3) how FEMA determined whether to recommend cost share adjustments, and their costs; and (4) FEMA's administrative cost percentages for declarations. GAO reviewed declaration data for fiscal years 2004-2011 and conducted site visits in 2011 to the two FEMA regions with the highest DRF obligations. The results are not generalizable, but provide insights.

In: FEMA's Disaster Declarations
Editors: M. Beck and C. Robbins
ISBN: 978-1-62808-188-6

Chapter 1

FEMA's Disaster Declaration Process: A Primer*

Francis X. McCarthy

Summary

The Robert T. Stafford Disaster Relief and Emergency Assistance Act (referred to as the Stafford Act - 42 U.S.C. 5721 et seq.) authorizes the President to issue "major disaster" or "emergency" declarations before or after catastrophes occur. Emergency declarations trigger aid that protects property, public health, and safety and lessens or averts the threat of an incident becoming a catastrophic event. A major disaster declaration, issued after catastrophes occur, constitutes broader authority for federal agencies to provide supplemental assistance to help state and local governments, families and individuals, and certain nonprofit organizations recover from the incident.

The end result of a presidential disaster declaration is well known, if not entirely understood. Various forms of assistance are provided, including aid to families and individuals for uninsured needs and assistance to state and local governments and certain non-profits in rebuilding or replacing damaged infrastructure.

The amount of assistance provided through presidential disaster declarations has exceeded $140 billion. Often, in recent years, Congress has enacted supplemental appropriations legislation to cover

* This is an edited, reformatted and augmented version of a Congressional Research Service publication, CRS Report for Congress RL34146, from www.crs.gov, dated May 18, 2011.

unanticipated costs. While the amounts spent by the federal government on different programs may be reported, and the progress of the recovery can be observed, much less is known about the process that initiates all of this activity. Yet, it is a process that has resulted in an average of more than one disaster declaration a week over the last decade.

The disaster declaration procedure is foremost a process that preserves the discretion of the governor to request assistance and the President to decide to grant, or not to grant, supplemental help. The process employs some measurable criteria in two broad areas: Individual Assistance that aids families and individuals and Public Assistance that is mainly for repairs to infrastructure. The criteria, however, also considers many other factors, in each category of assistance, that help decision makers assess the impact of an event on communities and states.

Under current law, the decision to issue a declaration rests solely with the President. Congress has no formal role, but has taken actions to adjust the terms of the process. For example, P.L. 109-295 established an advocate to help small states with the declaration process. More recently, Congress introduced legislation, H.R. 3377, that would direct FEMA to update some of its criteria for considering Individual Assistance declarations.

Congress continues to examine the process and has received some recommendations for improvements. Given the importance of the decision, and the size of the overall spending involved, hearings have been held to review the declaration process so as to ensure fairness and equity in the process and its results.

BACKGROUND

Under the Robert T. Stafford Disaster Relief and Emergency Assistance Act (P.L. 93-288) there are two principal forms of presidential action to authorize federal supplemental assistance. Emergency declarations are made to protect property and public health and safety and to lessen or avert the threat of a major disaster or catastrophe.[1] Emergency declarations are often made when a threat is recognized (such as the emergency declarations for Hurricane Katrina which were made prior to landfall) and are intended to supplement and coordinate local and state efforts prior to the event such as evacuations and protection of public assets. In contrast, a major disaster declaration is made as a result of the disaster or catastrophic event and constitutes a broader authority that helps states and local communities, as well as families and individuals, recover from the damage caused by the event.[2] The differences between the two forms of declarations remain an area of study regarding what events may or may not qualify for the respective declarations.[3]

Federal disaster assistance has served as the impetus for many supplemental appropriations bills over the last several decades, and has accounted for presidential declarations in every state and territory.[4] The supplemental funds are placed in the Disaster Relief Fund (DRF) which is a "no-year" fund managed by the Federal Emergency Management Agency (FEMA) and used only for spending related to presidentially declared disasters.

Major disasters can be a dominant story in the mass media that captures attention both for the devastation that results as well as the potential help that is expected. As one observer noted:

> Disaster assistance is an almost perfect political currency. It serves humanitarian purposes that only the cynical academic could question. It is largely funded out of supplemental appropriations and thus does not officially add to the budget deficit. It promotes the local economy of the area where the building process occurs.[5]

While disaster assistance may be good "political currency," a disaster declaration is generally the result of a tragic and devastating incident that disrupts (and sometimes takes) the lives of hundreds or thousands of families and individuals and the communities and states where they reside. The long-term economic and environmental impact of a disaster can be severe. The assistance offered from federal and private sources may or may not be commensurate with the damage inflicted by a natural or man-made event. Following a disaster, years of rebuilding and recovery work may lie ahead for communities and states. It is the declaration process that sets the federal recovery help in motion.

The trigger for federal disaster assistance is contained in a relatively short statutory provision. P.L. 93-288 (the Stafford Act) includes one brief section that establishes the legal requirements for a major disaster declaration:

> Section 401. Procedures for Declaration. All requests for a declaration by the President that a major disaster exists shall be made by the Governor of the affected state. Such a request shall be based on a finding that the disaster is of such severity and magnitude that effective response is beyond the capabilities of the state and the affected local governments and that the federal assistance is necessary. As a part of

> such request, and as a prerequisite to major disaster assistance under this Act, the Governor shall take appropriate response action under state law and direct execution of the state's emergency plan. The Governor shall furnish information on the nature and amount of State and local resources which have been or will be committed to alleviating the results of the disaster and shall certify that, for the current disaster, state and local government obligations and expenditures (of which state commitments must be a significant proportion) will comply with all applicable cost-sharing requirements of this Act. Based on the request of a Governor under this section, the President may declare under this Act that a major disaster or emergency exists.[6]

The process for an emergency declaration is also contained in the Stafford Act. In part it is similar to a major disaster declaration in finding and process, but the actual authorities are limited. In addition, section (b) provides for a special authority for the President to exercise his discretion for events that have a distinctly federal character:

> Sec. 5191. Procedure for declaration. (a) Request and declaration All requests for a declaration by the President that an emergency exists shall be made by the Governor of the affected State. Such a request shall be based on a finding that the situation is of such severity and magnitude that effective response is beyond the capabilities of the State and the affected local governments and that Federal assistance is necessary. As a part of such request, and as a prerequisite to emergency assistance under this chapter, the Governor shall take appropriate action under State law and direct execution of the State's emergency plan. The Governor shall furnish information describing the State and local efforts and resources which have been or will be used to alleviate the emergency, and will define the type and extent of Federal aid required. Based upon such Governor's request, the President may declare that an emergency exists. (b) Certain emergencies involving Federal primary responsibility. The President may exercise any authority vested in him by section 5192 of this title or section 5193 of this title with respect to an emergency when he determines that an emergency exists for which the primary responsibility for response rests with the United States because the emergency involves a subject area for which, under the Constitution or laws of the United States, the United States and authority. In determining whether or not such an emergency exists, the President shall consult the Governor of any affected State, if practicable. The President's

determination may be made without regard to subsection (a) of this section.[7]

The declaration process is elaborated upon in regulations, specifically in Subpart B of Part 206 (44 CFR). While these regulations have been adjusted through the regulatory process during the past three decades, since 1974 the procedures have undergone little significant change. The process itself is representative of the historical progression of federal disaster relief from being of an episodic nature to the current commonplace disaster declaration, now occurring on a weekly basis. The context in which disaster relief has grown has been in keeping with the growth of government and its concerns.

> Federal disaster relief has a long history in the U.S. dating back to the last years of the eighteenth century and arguably provided much of the political genesis for the New Deal social welfare programs (Landis 1999; Landis 1998; Moss 1999). As Michele Landis argues, social and political construction of claimants for relief as helpless victims of external forces beyond their control ("Acts of God") have exerted an enduring influence on American political discourse, which has manifested itself in heavy reliance on prior political precedents and analogies in constructing responses to current disasters.[8]

Information contained in the Preliminary Damage Assessments (PDAs) and summaries prepared by FEMA regional offices that accompany gubernatorial requests were long considered "pre-decisional and deliberative information" by the executive branch because they are part of the package that is developed and sent to the White House for the President's review and ultimate decision. These materials have generally not been available under the Freedom of Information Act process.[9] However, during 2008, FEMA began to list on its website the results of the PDAs. While the summaries of FEMA recommendations are still not available, the agreed upon figures from the PDAs are now available to the public for review.[10]

Policy makers have found it difficult to achieve equity in the treatment of disparate natural disaster events. The events can vary widely in their type, scope, duration, and impact. Perhaps the greatest variables are the states they affect. Each state has a different topography, a different history, and different capacities to respond and recover based on their own authorities, resources, and choices in what they will do following a disaster. Given those variations, it is a daunting task to construct a uniform process that can account for the range of natural and governmental circumstances that are a part of the nation's potential disaster landscape.

Congress and the Declaration Process

The Impetus for Reform

The question of how the federal disaster declaration process should work has come to the attention of Congress from time to time. Congress has requested reports from its investigative arms and investigated the process in panels that have considered aspects of the disaster relief process.[11] Various administrations have also considered the process and FEMA's role in it.

In 1981, and again in 2001, GAO issued reports on the declaration process that questioned the quality and consistency of FEMA's assessment criteria as well as the agency's ability to produce valid recommendations to the President on a governor's request for supplemental aid. As the most recent report (2001) concluded:

> These criteria are not necessarily indicative of a state's ability to pay for the damage because they do not consider the substantial differences in states' financial capacities to respond when disasters occur. As a result, federal funds may be provided for some disasters when they are not needed – a result that would be inconsistent with the Stafford Act's intent.[12]

Congressional interest in the declaration process derived, in part, from the increased cost in emergency spending—a recurring subject for the appropriations committees when considering supplemental spending legislation.[13] One review of disaster funding, particularly supplemental appropriations, found that the great majority of Disaster Relief Fund spending is related to the large, catastrophic events that meet the criteria established in the Stafford Act. These supplementals have been "driven" by the urgency of large natural disasters. As one witness explained in congressional testimony:

> Since funds provided in annual appropriations measures may be used to cover the federal costs of less catastrophic major disasters, Congress has focused critical attention on "must pass" legislation that would

replenish the Disaster Relief Fund for the worst and most costly disasters.[14]

A review of data for a seven-year period from 1988 to 1995 reveals that large expenditures, as funded by supplemental bills, relate to declarations issued for the largest events.[15] During this time period, disaster declarations were made for Hurricane Hugo, the Loma Prieta earthquake, Hurricane Andrew, the Midwest floods of 1993, and the Northridge earthquake. However, these were not the only events deemed worthy of presidential action and of cost to the federal treasury. As summarized by one author:

> But like the tail of a comet, over 200 other declarations accounted for one quarter of such outlays, many of them of relatively minute cost and extent. While of lesser impact on the national treasury, such "low end" declarations have become, to some observers, new sources of federal spending at the local level, long referred to in other contexts as "pork barrel spending."[16]

During early 2001, the term "entitlement" was beginning to be used in describing federal disaster spending by the new Bush Administration. Underlining this thinking in his first appearance before the Senate Appropriations Committee, FEMA Director Joe M. Allbaugh explained:

> FEMA is looking at ways to develop a meaningful and objective criteria for disaster declarations that can be applied consistently. These criteria will not preclude the President's discretion but will help states better understand when they can reasonably turn to the federal government for assistance and when it would be more appropriate for the state to handle the disaster itself.[17]

During the 110th Congress, attention was given to the disaster declaration process and what states and local governments can reasonably expect from that process. During the early spring of 2007, there were tornadoes that had an impact on Arkansas, Alabama and Georgia. The latter two states received disaster declarations while the damage in Arkansas was deemed insufficient to warrant federal assistance. Following these actions, Chairman Bennie Thompson of the House Homeland Security Committee scheduled a hearing to review FEMA's procedures. Representative Thompson set the context as follows:

> As Members representing real communities back home, we want to understand just how FEMA makes determinations regarding what is a disaster deserving of attention and when folks have to fend for themselves. Quite simply, we must have a serious discussion on what our expectations are of our federal government and what should remain a state and local responsibility.[18]

In order to examine the process and the financial projections of disaster spending in particular, Congress mandated in P.L. 110-28 that GAO study how FEMA develops its estimates for supplemental funding that may be needed.[19] Report language detailed these instructions for the review:

> The Committee continues to be concerned with FEMA's ability to manage resources in a manner that maximizes its ability to effectively and efficiently deal with disasters. One aspect of particular concern is how FEMA makes projections of funding needed in response to any given disaster or to meet future disasters. A recent Government Accountability Office (GAO) report raised concerns about FEMA's ability to manage its day-to-day resources and the lack of information on how FEMA's resources are aligned with its operations. As a follow-up to this report, the Committee requests that within six months of enactment GAO review how FEMA develops its estimates of the funds needed to respond to any given disaster. Such review should include how FEMA makes initial estimates, how FEMA refines those estimates within the first few months of a disaster, and how closely FEMA's estimates predict actual costs. The review should also include additional analysis and recommendations regarding FEMA's ability to manage disaster-related resources in a manner that maximizes effective execution of its mission.[20]

The study parameters mandated by Congress are broader than the declaration process, placing its emphasis on ongoing FEMA spending, including the refinement and revision of early estimates as well as the general management of spending projections for the Disaster Relief Fund (DRF). But it recognizes the importance of the initial estimates in decision-making and that the declaration process represents the fundamental decision to both establish federal participation following a disaster and provides the initial estimated amount of resources needed that informs that decision.

GAO's eventual report in response to this request suggested that FEMA's estimates for disaster costs have improved but could use additional refinement; for example

> some of the factors that can lead to changes in FEMA's cost estimates are beyond its control, such as the discovery of hidden damage. Others are not, such as its management of mission assignments. Sensitivity analyses to identify the marginal effect of key cost drivers could provide FEMA a way to isolate and mitigate the effect of these factors on its early estimates. To better predict applicant costs for the Individual Assistance program, FEMA could substitute or add more geographically specific indicators for its national average.[21]

During the 111th Congress, a hearing before the House Homeland Security Committee focusing on FEMA's regional offices also elicited comments regarding the relative speed, and perceived lack of transparency, of the declaration process. The comments came from both a committee member and a state official closely involved in the process.

> Ranking Republican Mike D. Rogers of Alabama questioned the timeliness of FEMA's disaster declarations. Declarations that should take a few days often take a month or longer, and a lack of available information on the status of requests only adds to state officials' frustration, said Brock Long, director of Alabama's Emergency Management Agency. "When we make phone calls to the region or to headquarters, a lot of time the answer we get is 'the declaration request is in process' and that's it," Long said. [22]

The Skepticism of Reform

There have been several proposals to impose more stringent regulations on the declaration process in an attempt to more precisely assess disaster impacts, calculate eligible damage, and incorporate some measure of suffering and loss. Differing perceptions of the declaration process resulted in different reactions to these proposed reforms.

The history of the disaster declaration process is rife with reform efforts that were perceived by some not as reform but as punitive measures directed at certain constituencies. Some of those differing perceptions were held by those closest to the event at the local level that had experienced a disaster. A different perception was also often held by governors that wanted to protect their option to request federal help. These perceptions were also shared, in some instances, by Members of Congress representing affected areas. All elected officials argued, in various forms, that reform should not impede the delivery of needed federal aid.

One example, now in law, of the desire to reform but not obstruct the declaration process is in the Post-Katrina Emergency Management Reform Act of October of 2006. That statute created a new position at FEMA: the Small State and Rural Advocate.[23] One of the principal duties of the advocate is to ensure that the needs of smaller states and rural communities will be "met in the declaration process;" the advocate is also directed to "help small states prepare declaration requests, among other duties."[24]

Another example was FEMA's response to recommendations from GAO's 1981 report. FEMA drafted regulations in 1986 that would be more certain in their delineation of state requirements prior to a declaration, reduce overall federal contributions, and install a formula to determine whether a state would receive a presidential disaster declaration for repairs to state and local infrastructure (known as "public assistance" in the Stafford Act). As the congressional report on the initiative summarized the FEMA draft:

> The proposal would have limited the number of future presidential declarations by establishing a "state deductible" based on a per capita minimum dollar amount adjusted by the ratio of the state/local price index to the national index. Of 111 declarations issued in prior years, 61 would have been ineligible for any public assistance. FEMA also proposed to decrease the federal share for disaster costs from 75 percent to 50 percent and to exclude aid to special districts.[25]

The FEMA proposal of 20 years ago addressed some problems identified by students and critics of the declaration process. However, some Members of Congress viewed this proposal as a means of removing the power of discretion from elected leadership. Rather than apply the empirical solution suggested by FEMA to perceived problems in the declaration process, Congress instead legislated a provision to explicitly forbid the primacy of any "arithmetic formula." In place of agreeing to the regulatory changes in the formula proposed by FEMA and the Administration, Congress added the following section to the Stafford Act:

> Limitation on the Use of Sliding Scales. Section 320. No geographic area shall be precluded from receiving assistance under this Act solely by virtue of an arithmetic formula or sliding scale based on income or population.[26]

While enactment of this legislation halted FEMA's efforts, the issue did not disappear. As discussed below, FEMA has since adopted regulations that use, but not "solely", arithmetic formulae in determining need for assistance.

PRESIDENTIAL AND GUBERNATORIAL DISCRETION

The declaration process contains many factors for consideration and, for all but the most catastrophic events, the process moves at a deliberate speed accumulating information from several sources. While the process is informed by that information and its relationship to potential assistance programs, the information that is gathered at the state and local level does not preclude the exercise of judgment by the governor or the President.

The Stafford Act stipulates several procedural actions a governor must take prior to requesting federal disaster assistance (including the execution within the state of the state emergency plan and an agreement to accept cost-share provisions and related information-sharing). Still, the process leaves broad discretion with the governor if he or she determines that a situation is "beyond the capabilities of the state." The concession that a state can no longer respond on its own is difficult to quantify. It is the governor who makes that assessment, based on his or her knowledge of state resources and capabilities.

Actions by a governor are a driving constant in this process. Both declarations of major disaster and declarations of emergency must be triggered by a request to the President from the governor of the affected state.[27] The President cannot issue either an emergency or a major disaster declaration without a gubernatorial request. The only exception to this rule is the authority given to the President to declare an emergency when "he determines that an emergency exists for which the primary responsibility for response rests with the United States because the emergency involves a subject area for which, under the Constitution or laws of the United States, the United States can exercise exclusive or preeminent responsibility and authority."[28]

The importance of governors in the process is not lost on, nor would it likely be diminished by, Presidents who formerly served in that position. Over the last three decades of Stafford Act implementation, four of the Presidents during this period were former governors who had worked through the disaster declaration process from both the state and the federal level.[29]

Having that experience may have left the Presidents, and their staffs and appointees, with an appreciation of the discretionary authority inherent in the process. While there are some established standards in the law, the factors that are to be weighed in considering the impact of a disaster on the need for assistance for families and individuals are general considerations that underscore the judgment required to reach a decision. As one observer noted of the decision for general, flexible considerations:

> In other words, Congress likes to keep the process imprecise, even if benefits occasionally go to the undeserving. The absence of objective criteria preserves wide political discretion to the president.30

Congress also has among its number former governors who have exercised this discretion at the state level. Still, despite the interest some may have in keeping "the process imprecise," some Members of Congress express disappointment at times with the exercise of the discretion and the general nature of the considerations. As noted previously in this report, that disappointment has been reflected in hearings that have focused on how the disaster declaration process works in practice.

In the declaration process, FEMA develops a recommendation that is sent to the White House for action. However, as implied, it is a recommendation from FEMA and the Department of Homeland Security (DHS). The final action, as defined in the regulations of the process, is a "Presidential determination."[31] Just as the governor retains the discretion to request federal assistance regardless of thresholds or indicators, the President retains the discretion to make a decision that may be counter to recommendations he receives.

Preliminary Damage Assessments

Although not explicitly mentioned in the Stafford Act, Preliminary Damage Assessments (PDAs) are a crucial part of the process of determining if an event may be declared a major disaster by the President. The minimal discussion of PDAs in the public record stands in inverse proportion to their impact on disaster decisions and subsequent expenditures from the Disaster Relief Fund (DRF).

When a PDA is conducted after an event it is the "mechanism used to determine the impact and magnitude of damage and the resulting unmet needs of individuals, businesses, the public sector, and the community as a whole."[32] The most "preliminary" part of a PDA may be an abbreviated one completed only by the state to determine if the situation merits development of a complete PDA with federal participation. Based on their previous experience, states may determine that the event will not reach the level where a federal disaster declaration is likely. However, despite findings that federal aid may not be needed, there may be political considerations that could lead to a gubernatorial request. As one author points out in his study of the process:

> Governors also feel the heat of media coverage of incidents in their states, and they too appreciate the importance of exhibiting political responsiveness. Governors also appreciate that their future political fortunes may be influenced by how they handle their disaster and emergency incidents. As a consequence, the hypothesis assumes that governors are the pivotal and decisive players in securing presidential disaster declarations and that they have a tendency to request declarations for even marginal events.[33]

While media and political pressure may have some influence on the outcome of some requests, governors may exercise caution since they are reluctant to be turned down when requesting aid. A denial of their request could be perceived by some to reflect adversely on their decision-making skills and judgment under pressure. Unlike the procedures of the federal process, a governor's decision to request a declaration is a public and often newsworthy action.

Regardless of any question regarding motivation, the governor's first decision is whether the incident is severe enough to assemble a traditional PDA team to survey the damaged area. The traditional PDA team includes a state official, representatives from the appropriate FEMA regional office, a local official familiar with the area and, in some instances, representatives from the American Red Cross and/or the Small Business Administration. The FEMA representatives have the responsibility of briefing the team on the factors to be considered, the information that will be helpful in the assessment and how the information should be reported. One significant improvement in this process is that the regulations now require that the participants reconcile any differences in their findings.[34]

Another factor is the quality of the PDA team and its findings. PDAs are ordered up quickly after an event. FEMA's 10 regional offices are often engaged in multiple disasters and have to rely on temporary employees (albeit, usually experienced ones) to staff the PDA team.[35] Also, given the variables among regions in interpreting policy and guidance, consistency of approach may also be a question. This became a significant issue during the hurricane season of 2004 in Florida, where questions were raised regarding the designations of some counties.[36]

When working openly and with federal and state cooperation, the PDA can be an effective and inclusive process. As Washington State Emergency Management Director Jim Mullen, and FEMA Region X Director Susan Reinertson, explained the process for a survey of flood damage in November of 2006:

> 'Joint PDA teams will visit and inspect damaged areas, document damage and talk, as needed, with homeowners and local officials,' said Mullen. 'It's a partnership effort designed to provide a clear picture of the extent and locations of damage in counties that have reported the most substantial damage to primary homes and businesses.'
>
> 'The PDA teams look at the total scope of damage to establish if recovery is beyond the capabilities and resources of the state and local governments. The PDA doesn't determine the total cost of recovery, nor does it guarantee a presidential declaration for individual assistance,' said FEMA Regional Director Susan Reinertson.[37]

PDA teams often face challenges in the collection of data. Some information may be observable in a survey of the area, such as the number of bridges damaged or the number of culverts washed out. But other necessary information, such as the percentage of elderly residents in an area, or the amount of insurance coverage for all homeowners or renters may be more difficult to obtain. Also the geographic span of the damage can create complications as the PDA team struggles to cover all of the affected area in a limited time. Further complications may then ensue based on how much of the area that the PDA teams have visited (generally counties) are included in the governor's request.

While PDAs are the usual way damages are assessed, there are exceptions to this rule. Some incidents are so massive in their scale and impact that the actual declaration is not in doubt. This would include events such as Hurricanes Andrew and Katrina, the Loma Prieta and Northridge earthquakes, the Mount St. Helens volcanic eruption, and the September 11 terrorist attacks. In these instances, the decision is not whether a declaration will be made, but how broad the coverage will be, both geographic and programmatic. In such cases the President, in accordance with the regulations, can waive the PDA requirement. But as the regulations note, a PDA may still be needed "to determine unmet needs for managerial response purposes."[38] This means the

PDA helps to identify a specific, potential need for certain programs, such as crisis counseling or disaster unemployment assistance during the disaster recovery period. It is this identification of discrete need that helps the governor decide on which assistance programs will be requested.

The PDA is a "bottom up" process as information gradually rises up for decision-makers to consider. Multiple pressure points, including affected citizens, elected officials, and professionals in various fields, may all urge PDA team members to reach certain conclusions. As one author suggests:

> The president, motivated by the need to appear highly politically responsive, solicits and encourages a gubernatorial request for a presidential disaster declaration. Publicity is a factor in that CNN and other news organizations help to promote nationally what would otherwise be a local incident addressed by subnational authorities. The president may also be influenced by the electoral importance of the state that experiences the incident.[39]

It is difficult to overstate the importance of the media context as noted above. Depending on the news of the day, the disaster event in question may be the biggest national story and thus create momentum for action that is difficult to assuage with explanations of traditional administrative procedures.

FACTORS CONSIDERED FOR PUBLIC ASSISTANCE IN MAJOR DISASTER DECLARATIONS

Public Assistance (PA) refers to various categories of assistance to state and local governments and non-profit organizations. Principally, PA covers the repairs or replacement of infrastructure (roads, bridges, public buildings, etc.) but also includes debris removal and emergency protective measures which cover additional costs for local public safety groups incurred by their actions in responding to the disaster.[40] In assessing the degree of PA damage, FEMA considers six general areas:

- Estimated cost of the assistance
- Localized impacts
- Insurance coverage
- Hazard mitigation

- Recent multiple disasters
- Programs of other federal assistance.

Estimated Cost of the Assistance

Although all of these factors are considered, the *estimated cost of the assistance* is a key component and may be "more equal" than other factors since it contains a threshold figure. What is especially noteworthy about the cost estimates is that they could be interpreted as an example of the "arithmetic formula" that was precluded from use in Section 320 of the Stafford Act. However, that section does state that a formula cannot be "solely" determinative of the fate of a governor's request. Use of the other factors listed arguably justifies FEMA's consideration of the threshold figure of $1 million in PA damage—the first number FEMA expects to see in a request that includes PA as an area of needed assistance. In addition FEMA also considers a state-wide threshold of $1.30 per capita in estimated eligible disaster costs before it will approve a request for PA help.[41]

Depending on the state's population, the per capita threshold may be difficult to reach. For example, the 2000 Census estimated California's population at just under 34 million people.[42] Applying the $1.30 per capita figure, it would require eligible PA damage in California to be close to $41.5 million. California is a large state with a budget and tax base commensurate with its size. Expecting a large state to be able to respond on its own is equating such help, and such amounts, to be within "the capabilities of the state."[43]

Compare that level of eligible damage for California ($41.5 million) with Nevada, a small population state according to FEMA regulations.[44] For Nevada, with a population of just under 2 million people according to the 2000 Census, eligible PA damage of about $2.4 million would make the state potentially eligible for supplemental federal assistance.

There are obvious differences in the populations of these two states; however, both have substantial industries and are growing areas. They are bordering states that are both subject to the threat of earthquake damage.[45] Different measurements have been suggested to try to more accurately capture a state's capacity to respond to disaster events. Some of these measurements are discussed in the "Congressional Considerations for the Declaration Process" section of this report.

Localized Impacts

The next factor used by FEMA to consider a declaration that may include PA help focuses on *localized impacts*. FEMA generally looks for a minimum of $3.27 per capita in infrastructure damage in a county before designating it for PA funding.[46]

While such a per capita amount would likely ensure that the local entity (almost always a county) would be included in the list of jurisdictions designated for PA assistance should a declaration be issued, high local levels of damage per capita would not supersede a finding that damages state-wide fell below the statewide threshold amount. However, knowledge of a very large localized impact could weigh on the President's discretion and raise the importance of the *localized impact* factor.

Insurance Coverage

Insurance coverage also is considered in assessing damage. Officials preparing PA estimates deduct the amount of insurance that should have been held by units of governments and nonprofit organizations from the total eligible damage amount. However, this is a complicated assessment with several caveats.

A considerable number of states and local governments "self-insure" their investments against some types of disasters and may argue that they would have total liability absent a federal disaster declaration.[47] Also, in the event of some disasters (earthquakes being a prime example), the state insurance commissioner must certify that hazard insurance is both available and affordable. If it is not available, deducting an amount of coverage not realistically available would not be a practical consideration.[48]

In the case of a flood event, it is much simpler for FEMA to access information available on whether flood insurance was available, given that FEMA administers the National Flood Insurance Program (NFIP) and there are program staff in each FEMA region who monitor NFIP participation. This knowledge of flood insurance availability and costs makes it a simpler process for FEMA to deduct the amount of flood insurance that should have been in place from the total potential award for a public structure.[49]

Hazard Mitigation

Hazard mitigation presents a challenging and different type of consideration for Public Assistance disaster declaration requests. If the requesting state can prove that their per capita amount of infrastructure damage falls short due to mitigation measures that lessened the disaster's impact, FEMA will consider that favorably in its recommendation to the President. Calculations of savings would be dependant on cost-benefit analysis and other related estimates of damages that were avoided. This factor is intended to encourage mitigation projects by states to lessen the risks of future natural disasters. Some might consider this practice to be one where "a good deed should go unpunished." However, another factor to be considered is that the mitigation work that is credited to the state may, in fact, have been principally financed (up to 75% of the costs) with previous FEMA disaster assistance funding through the Hazard Mitigation Grant Program (HMGP).[50]

Recent Multiple Disasters

Recent multiple disasters is the factor FEMA considers when a state has been repeatedly hit by disaster events (either presidential declarations or events within the state that were not declared) within the previous 12 months. FEMA evaluates the amount of funds that the state has committed to these recent events and their impact on the state and its residents. For example, a request from a state that has responded on it own to a series of tornadoes may receive a more favorable consideration, even if the catalyst for the request was arguably not as destructive as others. This factor was used recently in FEMA's assessment of the multiple hurricanes that struck Florida during the 2004 hurricane season.

Other Federal Programs

When FEMA is reviewing a governor's request it is also considering whether *other federal programs* are available. One example might be if a large amount of the reported damage occurred to federal-aid-system roads. The Federal Highway Administration (FHWA) is a more appropriate agency to handle such an event, and administers programs that address this specific type of damage.[51] The bridge collapse in Minnesota in August of 2007 was a dramatic example of this type of event that would warrant significant, non-FEMA, federal aid. Similarly, other federal programs may be more responsive to certain types of natural events and the problems they create. For example, oceanic bacteria could cause harmful failures for the fishing and hatcheries industry. Should such outbreaks occur, the Magnuson-Stevens Fisheries Act (P.L. 94-265) authorizes programs under the Commerce Department that would have more appropriate forms of emergency assistance for commercial fishermen than FEMA through the Stafford Act.[52] Also, while drought is a type of disaster that could result in a major disaster, such catastrophes usually result in program assistance from the U.S. Department of Agriculture rather than presidential declarations.[53]

In summary, all of these factors are used by FEMA to determine whether a major disaster declaration will be recommended and whether PA aid will be extended as a part of that potential declaration. FEMA's assistance to a homeowner who chooses not to purchase insurance is left with help from FEMA that is capped at $28,200. The assistance to repair public infrastructure is based on the amount of eligible damage caused by the disaster event. There is no cap, and the sums can be in the billions of dollars.

Factors Considered for Individual Assistance in Major Disaster Declarations

Individual Assistance (IA) includes various forms of help for families and individuals following a disaster event. The assistance authorized by the Stafford Act can include housing assistance, disaster unemployment assistance, crisis counseling and other programs intended to address the needs of people. In seeking to assess the impact of a disaster on families and individuals, the factors FEMA considers include:

- Concentration of damages
- Trauma
- Special populations
- Voluntary agency assistance
- Insurance coverage
- Average amount of Individual Assistance by state.

Concentration of Damages

Concentration of damages looks at the density of the damage in individual communities. FEMA's regulations state that highly concentrated damages "generally indicate a greater need for federal assistance than widespread and scattered damages throughout a state."[54] Concentrated damages are far more visible and may be an indication of significant damage to infrastructure supporting neighborhoods and communities, thereby increasing the needs of individuals and families.

However, the dispersion of damage is not necessarily an indication that individual and family needs are non-existent. Damage in rural states, almost by definition, is far less concentrated and could arguably be more difficult for a PDA team to view and assess. Congress has sought to address this challenge through the creation of the Rural and Small State advocate position at FEMA.[55]

Trauma

Trauma is defined in three ways in FEMA's regulations: the loss of life and injuries, the disruption of normal community functions, and emergency needs that could include an extended loss of power or water.[56]

Despite their prominence and importance to victims, families and communities, the loss of life and injuries have relatively little bearing on a declaration decision, but they would greatly influence media coverage that can influence the decision-making process.

An extreme amount of losses would be traumatic and considered in the evaluation of the governor's request. But the actual help available from FEMA in response to such losses is limited. The Other Needs Assistance (ONA) program, a part of the Individuals and Households Program (IHP), may provide assistance for uninsured funeral expenses and medical help.[57]

As noted, the extreme loss of life or many injuries will likely influence the amount of media coverage for the event. As previously explained in the "Presidential and Gubernatorial Discretion" section of this report, the media coverage can influence not merely the pace of the decision, but the actual decision itself.[58]

The other areas (disruption of functions and loss of utilities) in the trauma rubric can be a point of dispute in declaration decisions. What constitutes a disruption of normal community functions?

Do road closures that result in the closing of schools equate to a disruption? Another consideration is how long the disruption remains and how, or even if, federal help can alleviate the disruption. Similarly, the emergency needs due to the loss of utilities are defined by the length of time the power or water are not in service. Because characteristics of these factors are not defined in regulations, discretionary judgments are significant aspects of the evaluation of IA needs.

Special Populations

Special populations are considered by FEMA in assessing a request for Individual Assistance.

FEMA attempts to ascertain information about the demographics of an area affected by a disaster event. Those demographics include the age and income of residents, the amount of home ownership in an area, the effect on Native American tribal groups, and other related considerations to be taken into account. The knowledge of the demographics within the affected area gives FEMA added information to consider regarding trauma and community disruption. Special populations are a factor in the consideration of a governor's request, but the total number of households affected that would be eligible for Stafford Act programs remains the prime consideration.

Voluntary Agency Assistance

Voluntary agency assistance involves an evaluation of what the volunteer and charitable groups and state and local governments have already done to assist disaster victims as well as the potential help they can offer in the recovery period. FEMA also considers whether state or local programs "can meet the needs of disaster victims."[59] This factor is among the most contentious in the disaster declaration process due to the subjectivity of the assessment.

FEMA's evaluation of local and state capabilities, and the capabilities of the local voluntary community, may vary greatly among catastrophes. There is an expectation that the regional office's relationship and history with the state could provide some of this information. But a cursory look at state programs available to "meet the needs of disaster victims" suggests that few resources are comparable to federal help in intent or in scope. A National Emergency

Management Association (NEMA) survey of its members for state-funded disaster assistance showed that relatively few states provide assistance beyond that authorized in the Stafford Act.

NEMA noted that the "other" category could include a range of programs or funds not necessarily of direct aid to victims, including "local government loans, HAZMAT (hazardous materials) funds, disaster unemployment insurance, and a governor's contingency fund."[60] The report does not define these programs any further, but the small numbers of programs directed at IA or unmet needs suggests that FEMA officials evaluate this area carefully given the lack of information regarding available resources for these programs and the extent of their coverage.[61]

Table 1. Summarizes data from that survey

Assistance Provided	Number of States with These Programs
Public Assistance Program	22
Individual Assistance Program	9
Unmet Need Program	1
Other	5

Source: NEMA 2010 Biennial Report: Organizations & Funding for State Emergency Management and Homeland Security (Lexington, KY: Council of State Governments) p. 4.

In the same vein, attempts to assess the capacity of local voluntary and charitable groups to handle "unmet needs" caused by a disaster can be challenging and problematic. Part of the challenge is discerning the assistance available from the non-profit, voluntary sector, and if that aid meets the needs created by the disaster event. A problematic aspect of the assessment is similar to the "good deeds" concept addressed in the "hazard mitigation" factor in PA. In that instance, the criteria reward the mitigation work that communities and states had undertaken to lessen the impact of a disaster. Similarly, a strong and efficient charitable sector at the local level that is equipped and funded to address the remaining needs could result in a disaster not being declared and federal supplemental funding not being made available.

Arguably, a community does not base all of its preparedness decisions on the potential of FEMA funding in an extraordinary situation. However, it has been argued that some aspects of the FEMA declaration process could be viewed as disincentives for a sound, local capacity to deliver such assistance.

One analyst believed that this problem underlines the need for clear criteria for smaller disaster events:

> Too low a threshold reinforces the perception that the federal government will always come like the cavalry to rescue states and local governments from their improvident failure to prepare for routine disasters. Lapses in preparedness, response, recovery, and mitigation (to cite the disaster management litany) should not be encouraged by a too readily available bailout by the federal government and taxpayers.62

In evaluating the capabilities of local organizations, FEMA seeks to determine the help that actually exists, rather than assistance that might be expected to be in place.

Insurance Coverage

As noted earlier in the PA section, *insurance coverage* is also an important consideration when FEMA considers a request for Individual Assistance. This is in part derived from the general prohibition in the statute of the duplication of benefits, as follows:

> ... each federal agency administering any program providing financial assistance to persons, business concerns, or other entities suffering losses as a result of a major disaster or emergency, shall assure that no such person, business concern or other entity will receive such assistance with respect to any part of such loss as to which he has received financial assistance under any other program or from insurance or any other source.[63]

This provision does not necessarily result in delayed assistance. FEMA is able to provide help to individuals and households that have disaster damages but are waiting on insurance or other assistance for help. Those applicants can receive FEMA help as long as they agree to reimburse FEMA when they receive their other assistance. If a disaster occurred where almost all of the damaged dwellings were fully insured for the damage that was sustained, FEMA could conclude that a disaster declaration by the President was not necessary. Among the types of disasters FEMA frequently responds to, tornado disasters particularly reflect this challenge since tornado coverage is a part of most homeowners insurance policies.

Since the NFIP is administered by FEMA, officials can quickly determine the status of flood insurance in communities and the number of policies in place in the affected area. Additionally, knowledge of the income of the area's residents, as suggested in the *special populations* factor, allows the agency to make some projections regarding the likelihood of insurance coverage, particularly special hazard insurance such as flood or earthquake insurance, which are potentially expensive additions to a homeowners policy. As one analyst has noted:

> The decision not to buy insurance for earthquakes, floods, hurricanes and other natural disasters isn't always conscious: some homeowners don't realize they're not already covered. But many others, faced with high premiums and policies with limited coverage, gamble that they won't need insurance help to rebuild after a disaster.64

Average Amount of Individual Assistance by State

The last factor FEMA considers in assessing IA needs is the *average amount of Individual Assistance by state.* FEMA has issued statistics on average losses but notes that the average numbers used are not a threshold (see *Table 2*). The agency does suggest that the "following averages may prove useful to states and voluntary agencies as they develop plans and programs to meet the needs of disaster victims." The inference is that the levels listed generally are what would be expected in damage to dwellings.

The average numbers that follow are based on disasters that occurred between July of 1995 and July of 1999; the data are 10 years old and of questionable use today. The chart divides states into three categories: small states (under 2 million in population), medium states (2 to 10 million in population) and large states (over 10 million in population). The population amounts are based on the 1990 Census.

While FEMA's regulations stress that these are not thresholds, they are considered by agency officials in determining whether IA will be provided. Presumably states may consider that FEMA help could be forthcoming if damage reaches the state indicator levels. However, since the amounts have not been updated in ten years and are based on a 1990 census, it is difficult to determine the degree to which these numbers are considered. Given

Congress's mandate in Section 320 of the Stafford Act, this cannot be an arithmetic formula that solely determines whether assistance is provided. But the presentation of loss indicators may guide states when considering whether to request assistance.

In addition to the issue of the data's currency, a larger and more compelling question is whether the numbers in the chart match up to the Preliminary Damage Assessments (PDAs) for those events. Absent an existing review of detailed information in PDA forms, it is not possible to determine the usefulness of the data in *Table 2*.[65] Still, given the level of experience in this field following literally thousands of disaster declarations over the last 30 years, it could be argued that the numbers of eligible households assisted may be reflective of the estimated damage upon which the decision for a disaster declaration was made.

Table 2. Average Amount of Federal Assistance Per Disaster Based on Size of State, July 1995 - July 1999

Categories	Small States <2MillionPopulation	Medium States 2-10 Million Population	Large States >10 Million Population
Avg. population (1990 Census Data)	1,000,057	4,713,548	15,552,791
Number of disaster housing applications approved	1,507	2,747	4,679
Number of homes est. major damage/destroyed	173	582	801
Dollar amount of housing assistance	$2.8 million	$4.6 million	$9.5 million
Number of Individual and Family Grants (IFG) approved (Now known as ONA)	495	1,377	2,071
Dollar amount of IFG assistance (now ONA)	$1.1 million	$2.9 million	$4.6 million
Disaster housing/IFG (ONA) combined assistance	$3.9 million	$7.5 million	$ 14.1 million

Source: 44 CFR §206.48(b) (6).

Congressional Considerations for the Declaration Process

When Congress considers the mandated GAO reports and other commentary on the declaration process, there are some considerations that Members may wish to review when considering the current declaration process.

The Composition of Preliminary Damage Assessment Teams

One area of consideration is the composition of Preliminary Damage Assessment teams. Team members are involved throughout the process and include local officials guiding the team, state personnel who assist the governor in requesting assistance, and FEMA staff who work on the disaster if one is declared. While FEMA staff have the opportunity at several levels to refine the information the team gathers on damages and to ask additional questions, the process could be approached in other ways. Some FEMA staff have suggested the formation of several permanent teams that would have PDAs as their prime job task without continuing involvement in particular disasters. This could result in more consistent assessments with the bonus of added perspective of team members with exposure to various disasters in many regions of the country. Establishing independent and expert on-site groups to review a situation is a recommendation that the National Transportation Safety Board (NTSB) has been contemplating in their investigations of transportation accidents.[66]

The PDA is an important part of the declaration process. While it can be subject to challenge, it is also the assessment closest to the event. Given its vital role in the process the PDA deserves close attention to determine if these on-site assessments are accurately reflecting the character of an event and the likely eligible damages following a disaster event.

Updating and Revising Individual Assistance Averages

As previously noted, the averages that appear in FEMA's declaration process regulations for Individual Assistance (IA) to states are derived from experiences from July 1995 to July 1999.[67] These figures could be updated based on FEMA's more recent experience in delivering this type of assistance.

Legislation introduced in the 111th Congress had a provision that directed FEMA to review the factors that are considered for an IA designation.[68]

While the averages are separated by state population size, there does not appear to be a threshold number equivalent to the PA figure of $1 million in eligible damage. In the case of IA help, a dollar figure may not be desirable; however, numbers of families and households affected, or a minimum number of homes with major damage, could potentially be a starting point in establishing a base IA threshold for consideration of requests.

Other Potential Disaster Indicators

The current per capita indicator based on state population according to the U.S. Census is clear, but some observers believe it lacks precision. For example, the earlier discussion on PA indicators pointed out the commonality between California and Nevada regarding earthquake risk as well as the growth of the states. Their population sizes are very different but the per capita indicator alone does not necessarily measure a state's fiscal capacity.

The GAO report in 2001 noted that per capita personal income measures do not take into account the taxable income a state may enjoy from businesses or corporations that may have considerable taxable profits. GAO did suggest a different indicator:

> We have previously reported that Total Taxable Resources (TTR), a measure developed by the U.S. Department of Treasury, is a better measure of state funding capacity in that it provides a more comprehensive measure of the resources that are potentially subject to state taxation.[69]

FEMA's response to this suggestion questioned whether the use of TTR would be in violation of the legislative prohibition against arithmetic formulas.[70] The TTR does have very specific information and also tracks growth within a state.

However, this appears to make it a more accurate measurement. The degree of detail in the measurement would not necessarily make it the "sole" determinant of disaster aid. It could remain an indicator and one among several factors to be considered in the declaration process.

CONCLUDING OBSERVATIONS

The disaster declaration process, though subject to inquiry, argument, hearings, studies and recommendations, has changed very little over time. It remains a process that can be observed and evaluated as it occurs in the area affected by the disaster, and grows opaque as it moves up through layers of FEMA and DHS management to the White House. By making PDA information available, FEMA has begun to lift the veil on the decisions that are made. Congress has demonstrated an interest in this process and has sought to understand its limits and its effectiveness. "From the Government Accountability Office in 1981 to Vice President Gore's National Performance Review in 1994," one student of the process noted, "countless policy critiques have called for more objective criteria for presidential disaster declarations."[71] But those calls can be muted when a disaster occurs in a particular place that has particular importance to actors in the process, whether in Congress or the executive branch.

While criteria have been established to create a relatively uniform and knowable process, these criteria may not be determinative of the most critical elements considered. An emphasis on victims of unexpected natural disaster events will likely always have a compelling influence on the disaster declaration process. But since disaster relief does have a political element, at many levels, precedent arguably can have value in improving the declaration process so that it can be applied broadly and fairly. But precedent can also be problematic when discussing events of very different size and impact.

During the last Congress, attention was directed to the nature of the disaster declaration process. The 112th Congress may choose to consider a broad review of the process that might include the consistency of FEMA's approach across the nation in making damage assessments, other potential indicators of state capabilities and capacities, and the currency of the factors it employs to evaluate those assessments of disaster damage and the state requests on which they are based.

End Notes

[1] 42 U.S.C. § 5122.

[2] Ibid.

[3] For additional discussion, see CRS Report RL34724, *Would an Influenza Pandemic Qualify as a Major Disaster Under the Stafford Act?*, by Edward C. Liu.

[4] For more information on disaster supplementals, see CRS Report RL33226, *Emergency Supplemental Appropriations Legislation for Disaster Assistance: Summary Data*, by Justin Murray and Bruce R. Lindsay. For a map of disaster declarations, see http://www.bakerprojects.com.fema/Maps/CONUS.pdf.

[5] Rutherford H. Platt, *Disasters and Democracy: The Politics of Extreme Natural Events*, (Washington, DC: Island Press, 1999), p. 66.

[6] 42 U.S.C. § 5170.

[7] 42 U.S.C. § 5191.

[8] Michael J. Trebilcock and Ronald J. Daniels, "Rationales and Instruments for Government Intervention in Natural Disasters" in Ronald J. Daniels, Donald F. Kettl, and Howard Kunreuther, eds., *On Risk and Disaster: Lessons from Hurricane Katrina*, (Philadelphia: University of Pennsylvania Press, 2006), p. 104. For additional discussion of the evolution of emergency management see Claire B. Rubin, ed. *Emergency Management: The American Experience 1900-2005* (Fairfax, VA: Public Entity Risk Institute, 2007).

[9] U.S. Department of Homeland Security, Federal Emergency Management Agency, Annual Report of FOIA Activity for FY2000, II. C., at http://www.fema.gov/doc/help/foiareportfy00.doc.

[10] U.S. Department of Homeland Security, Federal Emergency Management Agency, *Preliminary Damage Assessment Reports*, at http://www.fema.gov/rebuild/recover/pda-reports.shtm.

[11] Ibid.; and Platt, p. 266.

[12] U.S. Government Accountability Office, *Disaster Assistance: Improvements Needed in Disaster Declaration Criteria and Eligibility Assurance Procedures*, GAO-01-837, August, 2001, p. 2.

[13] For more information on emergency supplemental spending for disasters see CRS Report RL33226, *Emergency Supplemental Appropriations Legislation for Disaster Assistance: Summary Data*, by Justin Murray and Bruce R. Lindsay.

[14] U.S. Congress, House Committee on the Budget, *Budgetary Treatment of Emergencies*, p. 68, 105th Congress, 2nd. Sess., June 23, 1998 (Washington: GPO, 1998).

[15] For more information, see CRS Report RL33053, *Federal Stafford Act Disaster Assistance: Presidential Declarations, Eligible Activities, and Funding*, by Keith Bea.

[16] Ibid.; and Platt, p. 10.

[17] U.S. Congress, Senate Committee on Appropriations, Subcommittee on VA, HUD and Independent Agencies, *Departments of Veterans Affairs and Housing and Urban Development and Independent Agencies Appropriations for Fiscal Year 2002*, 107th Congress, 1st sess., p. 252, at http://www.access.gpo.gov/congress/senate.

[18] Opening Statement of Rep. Bennie Thompson, in U.S. Congress, House Committee on Homeland Security, *Disaster Declarations, Where is FEMA in a Time of Need?*, 110th Congress, 1st. sess., March 15, 2007.

[19] P.L. 110-28, 121 Stat. 155.

[20] U.S. Congress, House Committee on Appropriations, *Making Emergency Supplemental Appropriations for the Fiscal Year ending September 30, 2007 and other purposes, report to accompany H.R. 1591*, 110th Cong., 1st sess., March 20, 2007, H.Rept. 110-60, p. 211.

[21] Government Accountability Office, *FEMA Disaster Cost Estimates*, Feb. 2008, p. 4.

[22] Matt Korade, "FEMA's Success Hinges on Regional Personnel," *Congressional Quarterly*, March 16, 2010, at http://homeland.cq.com/hs/display/do?docid=3569657&sourcetype=31.

[23] P.L. 109-295, 689g, 120 Stat. 1453.

[24] For more information see CRS Report RL33729, *Federal Emergency Management Policy Changes After Hurricane Katrina: A Summary of Statutory Provisions*, coordinated by Keith Bea.

[25] U.S. Congress, House Subcommittee on Investigations and Oversight, *The Federal Emergency Management Agency's Proposed Disaster Relief Regulations,* 100th Cong., 1st sess. (Washington: GPO, 1987), pp. 3-5.

[26] 42 U.S.C. 5163.

[27] For more information, see CRS Report RL33090, *Robert T. Stafford Disaster Relief and Emergency Assistance Act: Legal Requirements for Federal and State Roles in Declarations of an Emergency or a Major Disaster*, by Elizabeth B. Bazan.

[28] 42 U.S.C. §5191.

[29] President Carter (Georgia), President Reagan (California), President Clinton (Arkansas), and President George W. Bush (Texas).

[30] Ibid.; and Platt, p. 20.

[31] CFR 44 § 206.38.

[32] CFR 44 § 206.33.

[33] Richard Sylves, Quick Response Report #86: *The Politics and Administration of Presidential Disaster Declarations: The California Floods of Winter 1995,* at http://www.colorado.edu/hazards/research/qr/qr86.html.

[34] 44 CFR §206.33(c).

[35] FEMA has 10 regional offices: Region 1 (Boston, MA), Region 2 (New York, NY), Region 3 (Philadelphia, PA), Region 4 (Atlanta, GA), Region 5 (Chicago, IL), Region 6 (Denton, TX), Region 7 (Kansas City, MO), Region 8 (Denver, CO), Region 9 (Oakland, CA), and Region 10 (Bothell, WA).

[36] Sally Kestin and Megan O'Matz, "FEMA ruled on disaster before verifying Dade damage," *South Florida Sun-Sentinel*, at http://www.sun-sentinel.com/news/sfl-fema15may15,0,6848867.story?coll=sfla-news-utilities.

[37] U.S. Department of Homeland Security, Federal Emergency Management Agency, "Joint State-Federal Preliminary Damage Assessment Teams Visit Flood-Damaged Communities," Release Number R10-06-047, November 13, 2006, at http://www.fema.gov/news/newsrelease.fema?id=31509.

[38] 44 CFR 206.33(d)

[39] Ibid.; and Sylves, p. 5.

[40] U.S. Department of Homeland Security, Federal Emergency Management Agency, *Presidential Declarations: What Does This Mean for Me?,* at http://www.fema.gov/media/archives/2007/010807b.shtm.

[41] U.S. Department of Homeland Security, Federal Emergency Management Agency, "Notice of Adjustment of Statewide Per Capita Impact Indicator," 75 *Federal Register,* 62135, October 7, 2010.

[42] U.S. Census Bureau, *State & County Quick Facts,* at http://quickfacts.census.gov/gfd/states/06000.html.

[43] 42 U.S.C. §5170.

[44] 44 CFR § 206.48.

[45] U.S. Geological Survey, Earthquake Hazards Program, "Earthquake Density Maps for the United States", at http://earthquake.usgs.gov/regional/states/us_density.php.

[46] U.S. Department of Homeland Security, Federal Emergency Management Agency, "Notice of Adjustment of Countywide Per Capita Impact Indicator," 75 *Federal Register,* 62135, October 7, 2010.

[47] For information on public insurance coverage, see "Cost of Risk Survey" conducted by the Public Entity Risk Institute (PERI) at http://www.riskinstitute.org.

[48] 44 CFR §206.253. For additional information, see CRS Report RS22945, *Flood Insurance Requirements for Stafford Act Assistance*, by Edward C. Liu.

[49] 44 CFR §206.252.

[50] 42 U.S.C. § 5170c.

[51] For more information, see CRS Report RS22268, *Repairing and Reconstructing Disaster-Damaged Roads and Bridges: The Role of Federal-Aid Highway Assistance*, by Robert S. Kirk.

[52] For additional discussion of this law, see CRS Report RL34209, *Commercial Fishery Disaster Assistance*, by Harold F. Upton.

[53] For more information, see CRS Report RS21212, *Agricultural Disaster Assistance*, by Dennis A. Shields and Ralph M. Chite.

[54] 44 CFR §206.48(b)(1).

[55] P.L. 109-295, §689g, 120 Stat. 1453.

[56] 44 CFR §206.48(b)(2).

[57] 42 U.S.C. § 5174.

[58] Richard Sylves, Quick Response Report #86: *The Politics and Administration of Presidential Disaster Declarations: The California Floods of Winter 1995*, at http://www.colorado.edu/hazards/research/qr/qr86.html

[59] 44 CFR §206.48 (b) (4).

[60] National Emergency Management Association (NEMA) 2006 Biennial Report: Organizations and Funding for State Emergency Management and Homeland Security, (Lexington, KY: Council of State Governments).

[61] For more information on state assistance programs, refer to the following for summary information on state statutory authorities: CRS Report RL32287, *Emergency Management and Homeland Security Statutory Authorities in the States, District of Columbia, and Insular Areas: A Summary*, by Keith Bea, L. Cheryl Runyon, and Kae M. Warnock.

[62] Ibid.; and Platt, p. 65.

[63] 42 U.S.C. §5155.

[64] Liz Pulliam Weston, "Do You Really Need Disaster Insurance?," *MSN Money,* at http://moneycentral.msn.com/content/Insurance/Insureyourhome/P59648.hsp?Printer.

[65] This statement remains accurate. However, since FEMA has begun to list PDA information on its website, it should be possible to begin to assess the efficacy of the factors employed to determine IA damage and declaration criteria.

[66] For more information, see CRS Report RL33474, *Reauthorization of the National Transportation Safety Board (NTSB)*, by Bart Elias.

[67] 44 CFR § 206.48.

[68] H.R. 3377, Title III, Sec.304.

[69] U.S. Government Accountability Office, *DISASTER ASSISTANCE: Improvement Needed in Disaster Declaration Criteria and Eligibility Assurance Procedures,* GAO-01-837, August 2001, p.11.

[70] Ibid., p. 49.

[71] Ibid.; and Platt, p. 285.

In: FEMA's Disaster Declarations
Editors: M. Beck and C. Robbins
ISBN: 978-1-62808-188-6

Chapter 2

FEDERAL DISASTER ASSISTANCE: IMPROVED CRITERIA NEEDED TO ASSESS A JURISDICTION'S CAPABILITY TO RESPOND AND RECOVER ON ITS OWN*

United States Government Accountability Office

WHY GAO DID THIS STUDY

The growing number of disaster declarations—a record 98 in fiscal year 2011 compared with 65 in 2004—has contributed to increased federal disaster costs. FEMA leads federal efforts to respond to and recover from disasters and makes recommendations to the President, who decides whether to declare a disaster and increase the usual federal cost share of 75 percent. This report addresses (1) the number of declarations requested and approved from fiscal years 2004-2011 and associated DRF obligations; (2) the criteria FEMA used to recommend a declaration for PA, and the extent that FEMA assessed whether an effective response to a disaster was beyond the capabilities of state and local governments; (3) how FEMA determined whether to recommend cost share adjustments, and their costs; and (4) FEMA's administrative cost percentages for declarations. GAO reviewed declaration data for fiscal years

* This is an edited, reformatted and augmented version of United States Government Accountability Office, Publication No. GAO-12-838, dated September 2012.

2004-2011 and conducted site visits in 2011 to the two FEMA regions with the highest DRF obligations. The results are not generalizable, but provide insights.

WHAT GAO RECOMMENDS

GAO recommends, among other things, that FEMA develop a methodology to more accurately assess a jurisdiction's capability to respond to and recover from a disaster without federal assistance, develop criteria for 100 percent cost adjustments, and implement goals for and track administrative costs. FEMA concurred with the first two, but partially concurred with the third, saying it would conduct a review before taking additional action.

WHAT GAO FOUND

During fiscal years 2004-2011, the President received governors' requests for 629 disaster declarations and approved 539, or 86 percent, of which the Federal Emergency Management Agency (FEMA) reported 71 percent were for severe storms. For these 539 declarations, FEMA obligated $80.3 billion, or an average of about $10 billion a year, from the Disaster Relief Fund (DRF), as of January 31, 2012. Almost half of the obligations were for Hurricane Katrina; excluding obligations for Hurricane Katrina, FEMA obligated $40.6 billion, or an average of about $5 billion a year. As of January 31, 2012, FEMA anticipated that when all 539 declarations are closed, total DRF obligations will be about $91.5 billion.

GAO's analysis shows that FEMA primarily relied on a single criterion, the per capita damage indicator, to determine whether to recommend to the President that a jurisdiction receive public assistance (PA) funding. However, because FEMA's current per capita indicator, set at $1 in 1986, does not reflect the rise in (1) per capita personal income since it was created in 1986 or (2) inflation from 1986 to 1999, the indicator is artificially low. The indicator would be $3.57 in 2011 had it been adjusted for increases in per capita income and $2.07 in 2012 had it been adjusted for inflation from 1986 to 1999, rather than its current $1.35. GAO's analysis of FEMA's anticipated obligations for 508 declarations with PA during fiscal years 2004-2011 shows that 44 percent and 25 percent would not have met the indicator if it had been adjusted for increases in personal income and inflation, respectively, since 1986. Further, the per capita indicator does not accurately reflect a jurisdiction's capability to respond to or recover from a disaster without federal assistance. GAO identified other measures of fiscal capacity, such as total taxable resources, that could be more useful in determining a jurisdiction's ability to pay for damages to public structures. Developing a methodology to more comprehensively assess state capabilities and reexamining the basis for the indicator could help FEMA more accurately determine a jurisdiction's capacity to respond without federal assistance.

FEMA recommends raising the usual 75 percent federal share for PA to 90 percent when federal obligations, excluding FEMA administrative costs, meet a qualifying threshold. However, FEMA has no specific criteria for assessing requests to raise the federal share for emergency work to 100 percent, but relies on its professional judgment. For the 539 disaster declarations during fiscal years 2004-2011, governors made 150 requests to adjust the federal cost share to 90 or 100 percent; 109, or 73 percent, were approved or statutorily mandated, mostly for hurricanes. Without specific criteria for 100 percent cost share, FEMA risks making inconsistent or inequitable recommendations to the President.

GAO's analysis of administrative costs for 539 disaster declarations during fiscal years 2004-2011 shows that administrative cost percentages frequently exceeded FEMA's targets, although FEMA does not require that they be met. GAO's analysis of 1,221 disaster declarations shows that average administrative costs doubled from 9 to 18 percent during fiscal years 1989-2011, the time period for which FEMA has data available. FEMA is working on short- and long-term actions to improve efficiencies in delivering disaster assistance, but the agency does not plan to set goals or track performance for administrative costs. Until this happens, it will be difficult for FEMA to ensure assistance is being delivered in an efficient manner.

ABBREVIATIONS

DHS	Department of Homeland Security
DRF	Disaster Relief Fund
FCO	federal coordinating officer
FEMA	Federal Emergency Management Agency
GDP	gross domestic product
GSP	gross state product

IA	Individual Assistance
IFMIS	Integrated Financial Management Information System
JFO	joint field office
NEMIS	National Emergency Management Information System
PA	Public Assistance
PCI	state personal income
PDA	Preliminary Damage Assessment
PPD-8	Presidential Policy Directive-8
TTR	Total Taxable Resources

September 12, 2012
The Honorable Jason Chaffetz
Chairman
Subcommittee on National Security,
Homeland Defense, and Foreign Operations
Committee on Oversight and Government Reform
House of Representatives

The Honorable Roy Blunt
United States Senate

The growing number of major disaster declarations has contributed to an increase in federal expenditures for disaster assistance.[1] For example, during the 8 fiscal years from 2004 through 2011, Presidents declared 26 percent more major disaster declarations than during the preceding 8 fiscal years from 1996 through 2003. The Federal Emergency Management Agency (FEMA), a component of the Department of Homeland Security (DHS), leads the federal effort to mitigate, respond to, and recover from disasters, both natural and man-made. Disaster declarations can trigger a variety of federal response and recovery assistance for government and nongovernmental entities, households, and individuals. FEMA's disaster assistance programs include Individual Assistance (IA) and Public Assistance (PA), which provide financial assistance to individuals and jurisdictions, respectively.[2] Typically, FEMA pays 75 percent of the PA costs for disaster declarations, and state and local governments pay the other 25 percent. However, governors can request that the President approve an adjustment to the cost share, so that state and local governments would pay less than 25 percent.[3]

FEMA's Disaster Relief Fund (DRF) is the major source of federal disaster recovery assistance for state and local governments when a disaster is declared. The DRF is appropriated no-year funding, which allows FEMA to direct, coordinate, manage, and fund response and recovery efforts associated with domestic disasters and emergencies.[4] FEMA categorizes DRF obligations according to five categories: IA, PA, Hazard Mitigation, Mission Assignments, and Administration.[5] During fiscal years 2004 through 2011, PA was the category with the largest obligations. In addition, from fiscal years 2004 through 2011, the costs to administer some declarations exceeded the amount of federal assistance provided. In August 2011, the DRF diminished to a level that caused FEMA to temporarily halt funding on long-term recovery projects and focus on immediate needs. According to the FEMA Administrator, because of the shortage of available balances in the DRF, FEMA accelerated its efforts to recover previously obligated funds from states for completed projects that had unexpended balances. During fiscal year 2011, FEMA deobligated and recovered $2.2 billion from prior disaster declarations.

In 2001, we reported on the need for improvements to the criteria and eligibility procedures that FEMA used to make recommendations to the President for disaster declaration requests.[6] In 2012, we reiterated the need for FEMA to, among other things, reexamine the criteria FEMA primarily uses to make recommendations to the President for disaster declaration requests.[7] Furthermore, we have identified as a 21st century challenge the determination of costs to be borne by federal, state, and local governments or the private sector in preparing for, responding to, and recovering from disasters of all types.[8]

This report provides the results of our review of federal disaster assistance efforts by addressing the following questions: (1) For each fiscal year from 2004 through 2011, how many disaster declaration requests did FEMA receive, how many were approved, for which types of disasters, and how much were the associated obligations from the DRF? (2) What criteria has FEMA used to recommend to the President that a disaster declaration is warranted for PA, and to what extent does FEMA assess whether an effective response to a disaster is beyond the capabilities of state and local governments? (3) How does FEMA determine whether a cost share adjustment recommendation for PA is warranted and how much additional federal assistance did jurisdictions receive during fiscal years 2004 through 2011 because of cost share adjustments? (4) What were FEMA's administrative cost percentages for disaster declarations during fiscal years 2004 through 2011, how have they changed over time, and what actions is FEMA taking, if any, to reduce the costs of delivering disaster assistance?

To determine how many disaster declaration requests FEMA received, how many were approved, for which types of disasters, and how much the associated obligations were from the DRF, we obtained and analyzed data for each disaster declaration approved during fiscal years 2004 through 2011. We focused on this time frame because it contains the most current data for disaster declarations. It also comprises the time period after FEMA was merged into the newly created DHS, on March 1, 2003, and predates Hurricane Katrina in 2005. We focused primarily on fiscal years 2004 through 2011; however, to provide historical context and to compare results across similar periods, we also reviewed obligations data during fiscal years 1989 through 2011. In addition, to provide further historical perspective, we include information on the number of disaster declarations by jurisdiction from the first presidential disaster declaration in fiscal year 1953 through fiscal year 2011 in appendix I. FEMA provided data to us from its National Emergency Management Information System (NEMIS) and Integrated Financial Management Information System (IFMIS). To assess the reliability of these data, we reviewed the data that FEMA officials provided and discussed data quality control procedures with them. We determined that the data we used from these systems were sufficiently reliable for the purposes of this report.

To determine the criteria that FEMA used to recommend to the President that a disaster declaration was warranted for PA, and to what extent FEMA assessed whether an effective response to a disaster was beyond the capabilities of jurisdictions, such as state and local governments, we examined FEMA policies, regulations, and other documents related to the disaster declaration process. We also interviewed FEMA officials in headquarters and 2 of 10 FEMA regions as well as officials in two state emergency management agencies.[9] In addition, to determine the probability that a state was granted PA if the related damage estimate met or exceeded the PA per capita indicator, we obtained and analyzed data within Preliminary Damage Assessments (PDA) from fiscal years 2008 through 2011.[10] To assess the reliability of these data, we reviewed the data that FEMA officials provided and discussed data quality control procedures with them. We determined that the PDA data were sufficiently reliable for the purposes of this report.

To determine how FEMA evaluated whether a cost share adjustment recommendation was warranted and how much additional federal assistance jurisdictions received during fiscal years 2004 through 2011 because of the adjustments, we obtained and reviewed relevant laws, regulations, and policies. We also reviewed *Standards for Internal Control in the Federal Government* to assess FEMA's process for making recommendations to the

President on the need for cost share adjustments. 11 We also obtained and analyzed the cost share adjustments and types requested, approved, and denied during fiscal years 2004 through 2011. We interviewed FEMA officials who process cost share adjustment requests and participate in making recommendations to the President as to whether the requests should be approved or denied. To assess the reliability of these data, we reviewed the data that FEMA officials provided and discussed data quality control procedures with them. We determined that the cost share adjustment data were sufficiently reliable for the purposes of this report.

To determine what FEMA's administrative cost percentages were for disaster declarations, we obtained DRF obligations, projected obligations, and related data for 1,221 disaster declarations during fiscal years 1989 through 2011.[12] The projected obligations were FEMA's estimates, as of April 30, 2012, of the anticipated total obligations—actual to date and estimated—for each disaster eligible to receive federal disaster assistance after April 30, 2012.[13] We compared FEMA's administrative cost percentages for each disaster declaration from fiscal years 2004 through 2011 with FEMA's administrative cost percentage target ranges and FEMA's administrative cost percentages for disaster declarations from fiscal years 1989 through 2003.[14] We used these time periods because they provide larger sample sizes and an opportunity to assess potential trends over a longer period of time. In addition, we included projected obligations in our analysis because FEMA officials told us that administrative costs are typically higher as a percentage of assistance costs in the early months following a disaster. By including projected obligations, we were able to analyze disaster obligations—actual and estimated—for the entire time period that disaster declarations were in the following three levels of status: programmatically open, closed, and reconciled.[15] To assess the reliability of these data, we reviewed the data that FEMA officials provided and discussed data quality control procedures with them. We determined that the DRF data were sufficiently reliable for the purposes of this report. To determine what actions FEMA is taking, if any, to reduce the costs of delivering disaster assistance, we reviewed relevant FEMA documents and briefings, and interviewed FEMA officials.

We conducted this performance audit from July 2011 through September 2012 in accordance with generally accepted government auditing standards. [16] Those standards require that we plan and perform the audit to obtain sufficient, appropriate evidence to provide a reasonable basis for our findings and conclusions based on our audit objectives. We believe that the evidence obtained provides a reasonable basis for our findings and conclusions based on our audit objectives. For more information on our scope and methodology, see appendix II.

BACKGROUND

The Robert T. Stafford Disaster Relief and Emergency Assistance Act (Stafford Act), as amended, established the basic process for states to request a presidential disaster declaration.[17] The act also generally defines the federal government's role during the response and recovery after a disaster and establishes the programs and process through which the federal government provides disaster assistance to state and local governments, tribes, and certain nonprofit organizations and individuals.[18] In addition to its central role in recommending to the President whether to declare a disaster, FEMA has primary responsibility for coordinating the federal response when a disaster is declared as well as recovery, which typically consists of providing grants to assist state and local governments and certain private nonprofit organizations to alleviate the damage resulting from such disasters.

FEMA's disaster declarations process is implemented by FEMA headquarters as well as its 10 regional offices. FEMA's Administrator, in accordance with the Post-Katrina Emergency Management Reform Act of 2006 (Post-Katrina Act), appoints a Regional Administrator to head each regional office. [19] Regional Administrators—in partnership with state, local, and tribal governments, and other nongovernmental organizations— oversee emergency management activities within their respective geographical area.[20] Joint Field Offices (JFO) are temporary FEMA offices established to respond to declared disasters and are headed by Federal Coordinating Officers (FCO) who, among other things, coordinate the activities of the disaster reserve workforce deployed for a particular disaster. Once a disaster is declared, FEMA deploys Disaster Assistance Employees and any other employees needed to the affected jurisdiction(s).[21] FEMA provides assistance through the PA, IA, and Hazard Mitigation programs as well as through Mission Assignments. For instance, some declarations may provide grants only for IA and others only for PA. Hazard Mitigation grants, on the other hand, are available for all declarations if the affected area has a FEMA-approved Hazard Mitigation plan.

The Changing Disaster Declaration Environment

According to FEMA, the agency is evolving from originally focusing on grants management to being an organization implementing increasingly more complex programs, with an increasingly sophisticated and specialized workforce and procedures in response to changing circumstances and expectations. As illustrated by figure 1, the number of disaster declarations has significantly increased since 1953, when the first presidential disaster declaration was issued.[22] See appendix I for more information about the number of disaster declarations.

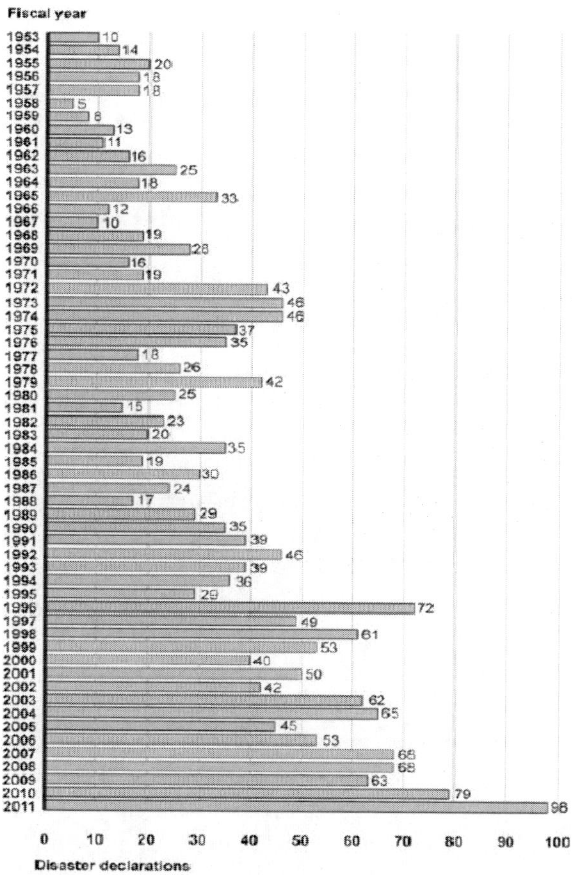

Source: GAO analysis of FEMA data.

Figure 1. Number of Disaster Declarations during Fiscal Years 1953 through 2011.

Various factors have contributed to the increase in disaster declarations. Population growth has occurred in U.S. geographic areas that are vulnerable when a disaster hits, such as those near coastlines. FEMA officials also cited more active weather patterns as a factor. FEMA guidance to states and localities and the enhanced capabilities and professionalization of state and local emergency management personnel have also been factors. For example, in 1999, FEMA published a list of factors that it considers when evaluating disaster declaration requests. According to FEMA and state emergency management officials from two states, the guidance, along with state and local emergency management officials' additional knowledge about the process and the enhanced transparency of the process for federal disaster declarations, has helped state and local officials better justify a request for federal disaster assistance. Increased media attention on disasters, especially those in which there have been casualties or deaths, has also been a factor, according to FEMA and state emergency management officials for two states.

Disaster Declaration Process

The disaster assistance process generally starts at the local level, proceeds to the county and state levels, and then to the federal level. The Stafford Act states that the governor of the affected state shall request a declaration by the President that a disaster exists.[23] FEMA is the primary federal disaster assistance agency, but others can have major roles, such as the U.S. Army Corps of Engineers, which can provide engineering and contracting support to FEMA.[24]

As part of the request to the President, a governor must affirm that the state's emergency plan has been implemented and the situation is of such severity and magnitude that effective response is beyond the capabilities of the state and the affected local governments and that federal assistance is necessary, among other things.[25] Before a governor asks for federal disaster assistance, state and local officials typically conduct an initial PDA to identify the amount of damage and determine if the damage exceeds their capability to respond and recover without federal assistance. Based on the initial PDA findings, a joint PDA, in which FEMA participates, may be requested by the governor.[26] FEMA uses the joint PDA in its evaluation of the state's need for federal assistance and makes a recommendation to the President as to whether the request for a disaster declaration should be approved or denied. Later in this report, we discuss in more detail how FEMA evaluates the need for PA.

To evaluate the need for IA, FEMA considers various factors, including insurance coverage; the extent to which volunteer agencies and state or local programs can meet the needs of disaster victims; concentration of damages due to the disaster; number of deaths and injuries; amount of disruption to normal community services; amount of emergency needs, such as extended or widespread loss of power or water; and special populations, such as elderly or low-income people.[27] Figure 2 shows the basic process that is followed from the time a disaster occurs until the President approves or denies a governor's disaster declaration request.[28]

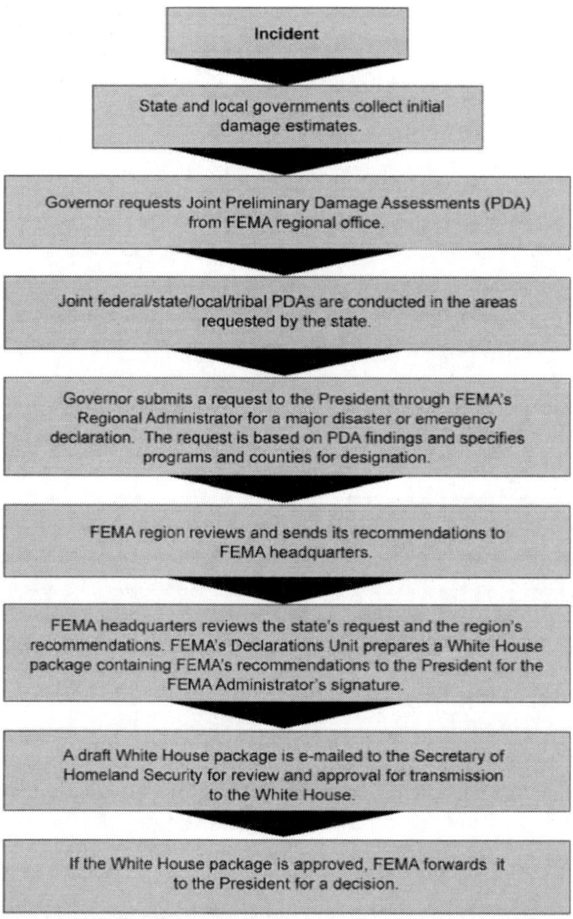

Source: GAO analysis of FEMA information.

Figure 2. Disaster Declaration Process.

OVER 500 DISASTERS WERE DECLARED DURING FISCAL YEARS 2004 THROUGH 2011 WITH TOTAL OBLIGATIONS OF OVER $90 BILLION ANTICIPATED

During this period, FEMA received 629 disaster declaration requests and approved 539 of them. Most disaster declarations were for severe storms. FEMA anticipates that when all disasters declared during fiscal years 2004 through 2011 are closed, its total obligations for these disasters will exceed $90 billion.

Most Declarations for Severe Storms, Highest in Southeast and Central Midwest

The President received requests from governors during fiscal years 2004 through 2011 for 629 disaster declarations and approved 539 of them, or 86 percent, as shown in table 1.

Table 1. Number of Disaster Declarations Requested and Approved during Fiscal Years 2004 through 2011

Fiscal year	Number requested	Number approved	Percentage approved
2004	76	65	86
2005	55	45[a]	82
2006	67	53	79
2007	78	68	87
2008	79	68	86
2009	72	63	88
2010	90	79	88
2011	112	98	88
Total	629	539	86

Source: GAO analysis of FEMA data.

[a]Despite the amount of destruction wrought by Hurricane Katrina in fiscal year 2005, only 4 of the 45 disaster declarations were associated with Hurricane Katrina—1 declaration each in Florida, Alabama, Mississippi, and Louisiana.

Governors can appeal a decision when the President initially denies a disaster declaration request.[29] During fiscal years 2004 through 2011, governors made 629 requests for disaster declarations, and the President ultimately denied 90 of them, or 14 percent.

FEMA has 10 regions throughout the United States that, among other things, provide technical assistance to state and local officials and make recommendations to FEMA headquarters as to whether a disaster declaration is warranted. Individual FEMA regions had varying numbers of disaster declarations during fiscal years 2004 through 2011. The two FEMA regions that had the most disaster declarations were Region IV in the Southeast and Region VII in the central Midwest, which together accounted for 163, or 30 percent, of the 539 declarations. The two FEMA regions that had the fewest declarations were Regions IX and X along the west coast, including Alaska, which together accounted for 71 declarations, or 13 percent. See appendix I, figure 10, for a map that shows the number of declarations by FEMA region.

During fiscal years 2004 through 2011, the average number of disaster declarations was 9.3 for each of the 58 jurisdictions—that is, the 50 states, the District of Columbia, 5 territories, and 2 Freely Associated States.[30] However, our analysis shows that some jurisdictions had over 20 disaster declarations, while other jurisdictions had 3 or fewer disaster declarations during this period. For example, Oklahoma had the most disaster declarations at 25, while Colorado and Guam had 1 each and the Marshall Islands did not have any. In addition, the 5 jurisdictions with the highest number of disaster declarations accounted for 105, or 19 percent, of the 539 declarations during fiscal years 2004 through 2011, whereas the 4 jurisdictions with the lowest number of disaster declarations accounted for 4, or less than 1 percent.[31] See figure 3 for the number of disaster declarations for each jurisdiction.

As reported by FEMA, severe storms accounted for 71 percent of declarations during fiscal years 2004 through 2011. According to FEMA officials, a disaster is classified as a severe storm when multiple storm-related incidents (for example, floods or heavy rains) affect a jurisdiction, but no single incident type is responsible for the majority of the damage.

See appendix I, table 11, for the number and percentage for each of the incident types that occurred during fiscal years 2004 through 2011.

For each disaster declaration, various types of assistance can be approved. For example, the President can approve PA only, IA only, or PA and IA for each declaration. As shown in figure 4, during fiscal years 2004 through 2011, 6 percent of the declarations were awarded for IA only, while a total of 94 percent of declarations were awarded for either PA only or IA and PA.

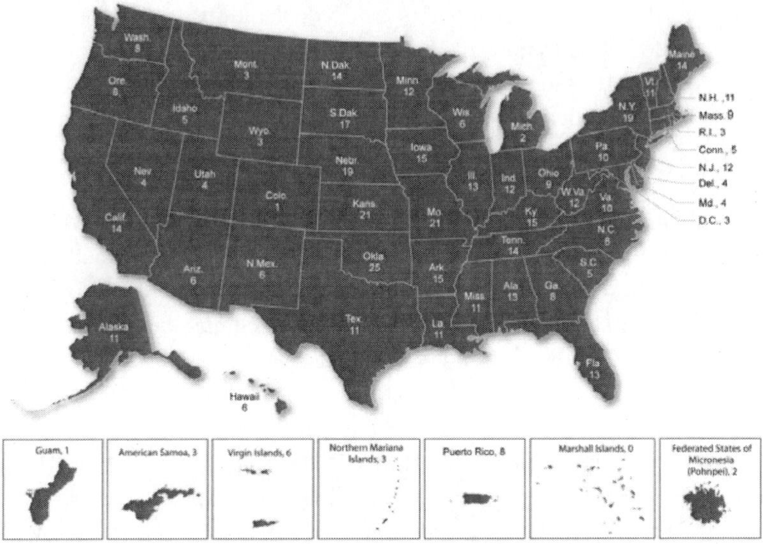

Source: GAO analysis of FEMA data, and Map Resources (map).

Note: In addition to Pohnpei, the Federated States of Micronesia consists of three additional major island groups, including Chuuk, Yap, and Kosrae.

Figure 3. Number of Disaster Declarations during Fiscal Years 2004 through 2011, by Jurisdiction.

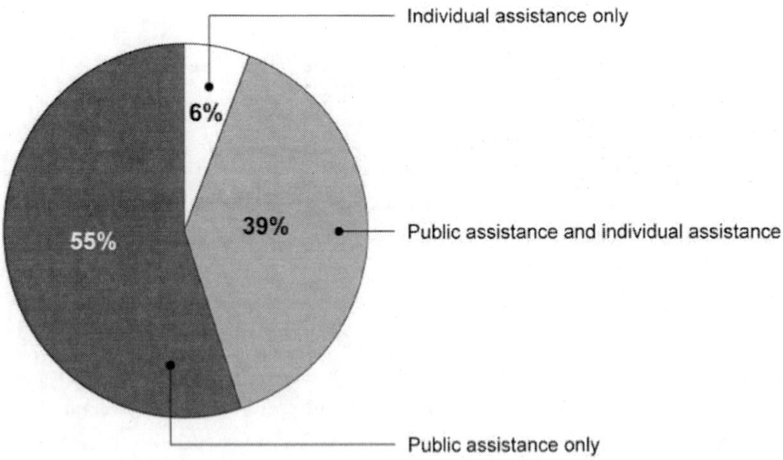

Source: GAO analysis of FEMA data.

Figure 4. Major Disaster Declarations by Type of Assistance, Fiscal Years 2004 through 2011.

Total Obligations of Over $90 Billion Anticipated for Disasters Declared during Fiscal Years 2004 through 2011

Through January 31, 2012, FEMA obligated $80.3 billion, or an average of about $10 billion a year, from the DRF for 539 disasters declared during fiscal years 2004 through 2011; and FEMA anticipates that when all 539 declarations are closed, obligations will be about $91.5 billion. Thirteen of these declarations had incurred obligations of over $1 billion each.[32] Almost half of the $80.3 billion in obligations was for Hurricane Katrina. Excluding obligations of $39.7 billion for Hurricane Katrina, FEMA obligated $40.6 billion for the other disaster declarations during fiscal years 2004 through 2011, or an average of about $5 billion a year. Total obligations are higher for fiscal years 2004, 2005, and 2008 than for the remaining 5 years primarily because of hurricanes that occurred with more frequency or force during those years.

For example, over half of the $8.8 billion for disasters declared in fiscal year 2004 was due to four hurricanes, over half of the $44.9 billion for disasters declared in fiscal year 2005 was for Hurricane Katrina, and about half of the $10.3 billion for disasters declared in fiscal year 2008 was for Hurricanes Ike and Gustav. Table 2 shows the obligations by fiscal year.

Obligations for disaster declarations during fiscal years 2004 through 2011 varied greatly by FEMA region and jurisdiction. FEMA Region VI had the highest obligations at $40.0 billion.

However, when excluding obligations from all FEMA regions due to Hurricane Katrina, FEMA Region IV had the highest obligations at $13.2 billion. FEMA Region X had the lowest obligations at $0.6 billion. As shown in figure 5, the amount of obligations also varied greatly by jurisdiction for disasters declared during fiscal years 2004 through 2011.

For example, Louisiana had the highest obligations, at $32.3 billion, but after excluding obligations for Hurricane Katrina, Florida had the highest obligations, at $9.3 billion, while for the jurisdictions with the lowest obligations, Guam had $1.9 million and the Marshall Islands did not have any. As a comparison, the nationwide average obligations per jurisdiction were $1.38 billion, and decreased to $700 million when obligations for Hurricane Katrina were excluded.

Appendix I, figure 10, shows the obligations by FEMA region for fiscal years 2004 through 2011.

Table 2. Obligations by Fiscal Year for 539 Disaster Declarations during Fiscal Years 2004 through 2011 (Dollars in millions)

Fiscal year	Public Assistance	Individual Assistance	Hazard Mitigation	Mission Assignment	Administration	Total[a]
2004	$3,725	$2,741	$435	$817	$1,077	$8,794
2005 (including Katrina)	17,770	15,529	1,431	4,143	5,993	44,867
2005 (excluding Katrina)	2,067	1,504	278	472	776	5,097
2006	2,605	633	206	151	386	3,982
2007	1,628	336	238	6	281	2,488
2008	6,116	1,880	816	157	1,394	10,364
2009	1,700	340	236	41	455	2,772
2010	1,616	828	120	6	451	3,022
2011	1,599	1,213	41	438	769	4,061
Total (including Katrina)	$36,759	$23,500	$3,522	$5,761	$10,806	$80,349[a]
Total (excluding Katrina)	$21,057	$9,476	$2,369	$2,090	$5,589	$40,579[a]

Source: GAO analysis of FEMA data.

Note: This analysis shows obligations from the DRF only and does not include any potential administrative cost expenditures from FEMA's other annual appropriations. Also, the obligations exclude obligations by other federal agencies and state and local governments.

[a]Obligations for each year may not add up to the total obligations reported because of rounding.

Furthermore, obligations for individual disaster declarations declared during fiscal years 2004 through 2011 varied greatly. For example, as of January 31, 2012, FEMA had obligated $28.5 billion for Louisiana's fiscal year 2005 Hurricane Katrina disaster declaration compared with about $803,000 for a South Dakota disaster declaration during that same fiscal year. Disaster declarations can take over a decade to close; therefore, to obtain a more comprehensive and longer-term perspective, we analyzed obligations for 811 disaster declarations during fiscal years 1989 through 2011 that had been closed as of January 31, 2012.[33] Of the 811 closed declarations, we found that 440, or 54 percent, had obligations of less than $10 million (see table 3).

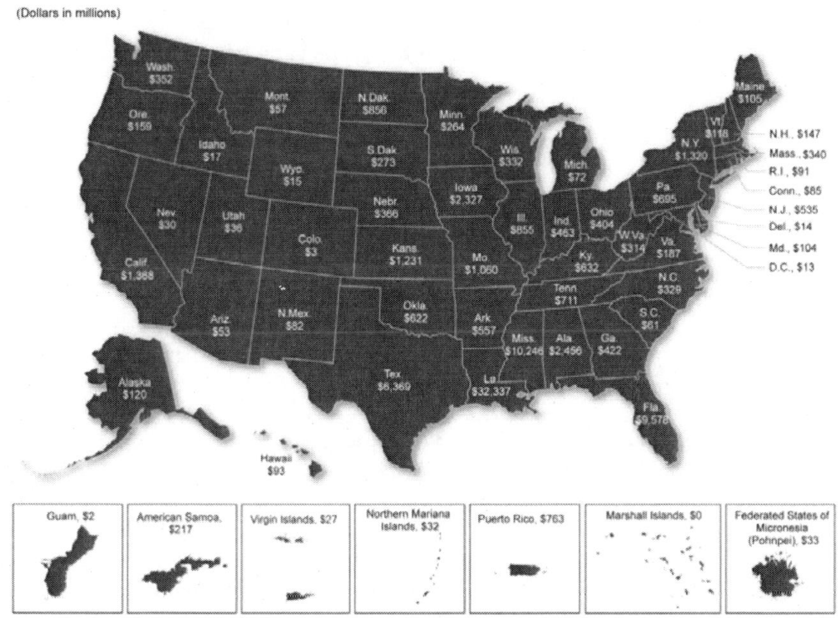

Source: GAO analysis of FEMA data, and Map Resources (map).

Note: For the jurisdictions that had disaster declarations for Hurricane Katrina, the obligations excluding Hurricane Katrina were Alabama, $1.4 billion; Florida, $9.3 billion; Louisiana, $3.8 billion; and Mississippi, $236 million. In addition to Pohnpei, the Federated States of Micronesia consists of three additional major island groups, including Chuuk, Yap, and Kosrae.

Figure 5. Disaster Relief Fund Obligations, Including Hurricane Katrina, by Jurisdiction during Fiscal Years 2004 through 2011, as of January 31, 2012.

For those disaster declarations approved during fiscal years 2004 through 2011, we analyzed the total obligations as of April 30, 2012, for closed disasters, and the total projected obligations—actual to date and estimated—for those declarations that remained open as of April 30, 2012.[34] Specifically, for open declarations as of April 30, 2012, instead of analyzing how much FEMA had obligated as of that date, we analyzed the amount FEMA had obligated plus the amount FEMA anticipated it would obligate from the time a declaration was approved through its closure. On the basis of our analysis, when all 539 declarations that were declared during fiscal years 2004 through 2011 are eventually closed, FEMA anticipates that 193, or 36 percent, will have total obligations of less than $10 million, thus signifying that these were relatively small disasters (see table 4).

Table 3. Number and Percentage of Closed Disasters Declared during Fiscal Years 1989 through 2011, by Obligation Amounts as of January 31, 2012

Obligations for each disaster declaration[a]	Number of disaster declarations	Percentage of total
Less than $1 million	21	3
$1 million to less than $5 million	261	32
$5 million to less than $10 million	158	19
$10 million to less than $25 million	179	22
$25 million to less than $50 million	81	10
$50 million to less than $100 million	55	7
At least $100 million	56	7
Total	811	100

Source: GAO analysis of FEMA data.

[a]Obligations include the funds obligated by FEMA from the DRF and exclude obligations by other federal agencies and state and local governments. These amounts are not in constant, inflation-adjusted dollars, because the data were not detailed enough to identify the specific year an obligation was made.

Table 4. Number and Percentage of Disasters Declared during Fiscal Years 2004 through 2011, by Combined Actual and Projected Total Obligation Amounts, as of April 30, 2012

Obligations for each disasterdeclarationa	Number of disaster declarations	Percentage of total
Less than $1 million	5	1
$1 million to less than $5 million	99	18
$5 million to less than $10 million	89	17
$10 million to less than $25 million	140	26
$25 million to less than $50 million	76	14
$50 million to less than $100 million	55	10
At least $100 million	75	14
Total	539	100

Source: GAO analysis of FEMA data.

[a]Obligations include the funds obligated by FEMA from the DRF and exclude obligations by other federal agencies and state and local governments. These amounts are not in constant, inflation-adjusted dollars, because the data were not detailed enough to identify the specific year an obligation was made.

ELIGIBILITY FOR ASSISTANCE IS PRIMARILY DETERMINED USING A DAMAGE ESTIMATE INDICATOR

The per capita damage indicator FEMA uses to assess a jurisdiction's eligibility for PA is the primary factor on which disaster declaration decisions are based. However, the per capita damage indicator is artificially low. In addition, FEMA's process to determine eligibility for federal assistance does not comprehensively assess a jurisdiction's capability to respond to and recover from a disaster on its own.

FEMA Relies Almost Exclusively on a Single Indicator

According to FEMA and state emergency management officials, FEMA has primarily relied on a single indicator, the statewide per capita damage indicator, to determine whether to recommend that a jurisdiction receive PA funding. In fiscal year 2012, the per capita indicator is $1.35.[35] Thus, a state with 10 million people would generally have to incur $13.5 million in estimated eligible disaster damages to public structures for FEMA to recommend that a disaster declaration for PA is warranted.[36] However, other factors could also influence the recommendation, such as whether a jurisdiction has incurred multiple disasters within a short period of time. Of the 58 jurisdictions for fiscal year 2012, based on population, California has the highest statewide indicator total, at $50.3 million, while Wyoming has the lowest amount at $760,895.[37] See appendix III, table 15, for the total PA per capita indicator amounts for each of the 58 jurisdictions.

FEMA's method to determine the affected jurisdictions' capabilities to respond without federal assistance relies on a governor's certification and damage estimates. The Stafford Act requires that a governor's request for a disaster declaration be based on a finding that the disaster is of such severity and magnitude that an effective response is beyond the capabilities of the jurisdiction and that federal assistance is necessary. 38 FEMA officials stated that governors must certify in their letter to the President requesting a disaster declaration that the disaster is beyond the capabilities of the jurisdiction. FEMA regulations list quantitative and qualitative factors, such as recent disasters within the same jurisdiction that the agency considers when determining whether a disaster declaration is warranted.[39] However, in describing the declarations process, FEMA and emergency management officials in two states said that FEMA uses the statewide per capita indicator as the primary determining factor for PA funding. This damage indicator, which FEMA has used since 1986, is essentially a proxy fiscal measure of a state's capacity to respond to and recover from a disaster, rather than a more comprehensive assessment of a state's fiscal capacity.

According to our analysis of readily available indicator data, as well as officials in two FEMA regions and state emergency management officials in two states, the principal factor used to determine eligibility for a disaster declaration was whether the damage estimate exceeded the PA per capita indicator. Our analysis of 246 disaster declarations during fiscal years 2008 through 2011 identified the PA per capita indicator as having been the primary determining factor—essentially being used as an eligibility threshold.[40] Specifically, 244 of the 246 approved disaster declarations that we reviewed, or 99 percent, had PA damage estimates that met or exceeded the PA per capita indicator in effect in the year in which the disaster was declared.[41] Seven gubernatorial requests for a disaster declaration during fiscal years 2008 through 2011 had a damage estimate higher than the PA per capita indicator yet were denied for various reasons, such as the damage being a result of multiple storms or the normal depreciation of structures rather than a single disaster.

FEMA's Per Capita Indicator Used to Assess Eligibility Is Artificially Low

Because FEMA's current per capita indicator does not reflect the rise in (1) per capita personal income since it was created in 1986 or (2) inflation from 1986 to 1999, the indicator is artificially low. In 1986, FEMA proposed a $1.00 per capita indicator for PA as a means of gauging a jurisdiction's fiscal capacity.[42] The indicator was based on the 1983 per capita personal income nationwide, then estimated at $11,667. Current FEMA officials were unable to explain how per capita personal income was used to establish the indicator level at $1.00. However, FEMA documentation noted that the agency thought it reasonable that a state would be capable of providing $1.00 for each resident of that state to cover the damage from a disaster. While the proposed rule was not codified in 1986, FEMA began to use the $1.00 per capita indicator informally as part of its preliminary damage assessment efforts and did not adjust the indicator annually for either inflation or increases in national per capita income. In 1998, FEMA considered adjusting the PA indicator to $1.51

to account for inflation since 1986, but because of input from state emergency management officials, FEMA decided not to do so.

In 1999, FEMA issued a rule codifying the per capita indicator at $1.00, which was stipulated to include an annual adjustment for inflation, but the rule was silent on whether the indicator would continue to be based on nationwide per capita personal income.[43] As a result, the indicator has risen 35 percent from $1.00 to $1.35 in the 13 years since FEMA began its annual inflationary adjustments. Figure 6 shows the actual increases in the per capita indicator for PA from 1986 to 2010 compared with the increases that would have occurred if FEMA had adjusted the indicator for inflation or the increase in per capita personal income during this period.

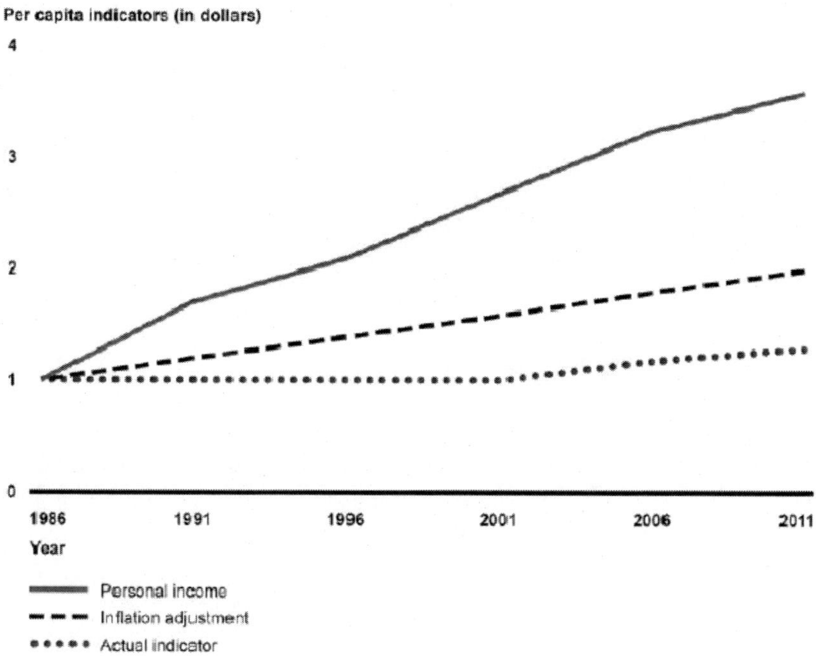

Source: GAO analysis of Department of Homeland Security, Department of Commerce, and Bureau of Labor Statistics data.

Figure 6. Comparison of Actual Public Assistance Per Capita Indicator with the Indicators if They Had Been Adjusted for Increases in Personal Income and Inflation, 1986 through 2011.

FEMA officials stated that the rise in construction and other costs to respond to and recover from disasters have outpaced the rise in the per capita indicator. In jurisdictions with smaller populations, damage to a single building or facility, such as a water treatment facility, could result in a damage estimate sufficient to meet the per capita damage threshold and warrant a disaster declaration. For example, the damage from Hurricane Katrina to a single water treatment facility in Carrollton, Louisiana, exceeded Louisiana's 2005 per capita threshold.[44] In addition, the Washington National Cathedral incurred approximately $15 million of damage during the August 23, 2011, earthquake in Washington, D.C., which has a per capita damage indicator threshold of less than $1 million.[45]

The Stafford Act requires that conditions due to a disaster be beyond a jurisdiction's (state and local) capability to respond effectively before disaster assistance from the federal government is warranted.[46] The law, however, prohibits FEMA from denying federal assistance solely by virtue of an arithmetic formula or sliding scale based on income or population.[47] According to *Standards for Internal Control in the Federal Government*, activities should be established to monitor indicators and controls should be aimed at validating the propriety and integrity of such indicators.[48] Had the indicator been adjusted for inflation beginning when FEMA started using it in 1986, the indicator would have risen more than 100 percent, from $1.00 to $2.07 in 2012. Had the indicator been adjusted for increases in per capita personal income since 1986, the indicator would have risen over 250 percent, from $1.00 to $3.57 in 2011, based on 2011 national per capita personal income of $41,663.

Our analysis of actual and projected obligations for 508 disaster declarations in which PA was awarded during fiscal years 2004 through 2011 showed that fewer disasters would have met either the personal income-adjusted or the inflation-adjusted PA per capita indicators for the years in which the disaster was declared. 49 Specifically, our analysis showed that 44 percent of the 508 disaster declarations would not have met the PA per capita indicator if adjusted for the change in per capita personal income since 1986. Similarly, our analysis showed that 25 percent of the 508 disaster declarations would not have met the PA per capita indicator if adjusted for inflation since 1986.[50] Thus, had the indicator been adjusted annually since 1986 for personal income or inflation, fewer jurisdictions would have met the eligibility criteria that FEMA primarily used to determine whether federal assistance should be provided, which would have likely resulted in fewer disaster declarations.[51]

In discussions with FEMA officials about raising the per capita damage indicator, they noted that updating the indicator completely in a single year could create problems for jurisdictions, which, in response, may need to increase their rainy day fund or take other actions to adjust to the change. However, FEMA officials stated that adjusting the indicator in a phased approach over several years would be more feasible for jurisdictions. The current annual inflation adjustment generally increases the damage indicator incrementally. However, were the "catch-up" inflation adjustment (from $1.35 to $2.07) implemented in a single year, the increase would be considerably more than the annual inflation adjustments since 1999. For example, for a jurisdiction with a population of 5 million, fully implementing the catch-up adjustment for inflation would raise the damage indicator from $6.75 million to $10.35 million. Adjusting the indicator in phases over several years could help FEMA examine future requests for disaster declarations in a manner that reflects changes in per capita income or inflation since 1986 and provide jurisdictions more time to plan for and adjust to the change.

FEMA's Eligibility Process Does Not Comprehensively Assess a Jurisdiction's Capability to Respond and Recover

Reliance on the PA per capita indicator to determine a jurisdiction's eligibility for federal assistance—whether the indicator is artificially low or adjusted for increases in personal income or inflation—does not provide an accurate measure of a jurisdiction's capability to respond to or recover from a disaster without federal assistance. Determining a jurisdiction's fiscal capacity is important because a jurisdiction with greater resources should be able to more easily recover in the aftermath of a disaster than a jurisdiction with fewer resources. Further, a jurisdiction's fiscal capacity is an important component of the jurisdiction's overall response and recovery capability.

In 1999, when the rule was codified to set the per capita indicator at $1.00, FEMA stated that it recognized that a straight per capita figure may not be the best measurement of a state's capability, but that it provided a simple, clear, consistent, and long-standing means of measuring the severity, magnitude, and impact of a disaster while at the same time ensuring that the President can respond quickly and effectively to a governor's request for assistance.[52] As we reported in 2001, per capita personal income is a relatively poor indicator of a jurisdiction's fiscal capacity because it does not comprehensively measure all income potentially subject to jurisdiction taxation and is not necessarily indicative of jurisdiction or local capability to respond effectively without federal assistance.[53] For example, it does not include income produced in a jurisdiction unless it is received as income by a jurisdiction resident. Thus, profits retained by corporations for business investment, though potentially subject to jurisdiction taxation, are not included in a jurisdiction per capita income measure because they do not represent income received by jurisdiction residents. In 2001, we recommended that FEMA consider alternative criteria. FEMA's response noted that we provided valuable input for the FEMA team that was reviewing the disaster declaration process and the criteria used to assess jurisdiction damages. According to FEMA, in 2001, the President's budget for fiscal year 2002 included a provision for the development of improved guidelines for disaster assistance that provided jurisdictions with meaningful criteria that must be met to become eligible for federal disaster assistance. FEMA undertook a review of disaster declaration guidelines; however, no changes to the established declaration guidelines were adopted, and ultimately, FEMA did not change its reliance on the per capita indicator.

The Post-Katrina Act required FEMA to develop a set of preparedness metrics that could be used to assess operational preparedness capability.[54] Also, Presidential Policy Directive-8 (PPD-8), issued in March 2011, required the Secretary of Homeland Security to develop a national preparedness system to, in part, define existing capabilities and capability gaps, and drive investments to close those gaps across the nation's federal, state, local, tribal, and territorial governments. Much of the growth in disaster declarations has occurred at the same time (that is, since the terrorist attacks of September 11, 2001) that the federal government has provided more than $37 billion to state and local governments to enhance their preparedness to protect against, respond to, and recover from disasters of all types. However, FEMA has not yet finished developing metrics to assess state preparedness capability, a fact that limits its ability to comprehensively assess jurisdictions' disaster preparedness and capabilities.

Without an established means of assessing jurisdiction response and recovery capability, FEMA has continued to rely primarily on the per capita damage indicator when determining whether a disaster declaration is warranted. The National Preparedness Goal, released in September 2011, identifies 31 core capabilities and targets. Further, FEMA officials stated that jurisdictions provided reports to FEMA on their preparedness, including their core capabilities, on December 31, 2011. According to FEMA, the state preparedness reports resulted in up to 155 individual measures to track residual capability gaps, the results of which are reported by core capability in the annual *National Preparedness Report*. These core capabilities are the latest evolution of the Target Capabilities List.[55]

While the preliminary core capability targets provide a basis for jurisdictions to understand the core capabilities, FEMA has not yet developed national preparedness capability requirements based on established metrics for the core capabilities at the state and local levels— as called for in the Post-Katrina Act and PPD-8. In addition, according to FEMA officials, FEMA does not have any plans or policies in place to use preparedness data to inform its recommendations regarding presidential disaster declarations. Metrics to assess a jurisdiction's disaster preparedness and capabilities could augment the PA per capita indicator, and other relevant information, to provide a more comprehensive understanding of a jurisdiction's capacity to respond to and recover from a disaster without federal assistance.

The 2011 state preparedness reports provide some potentially useful information to understand a state's response capabilities. However, FEMA does not use these reports or an assessment of a jurisdiction's response capabilities to determine eligibility for disaster assistance, and the FEMA Administrator stated that state and local governments are capable of handling much of the workload related to responding to a declared disaster, which has allowed FEMA to mostly focus on recovery efforts.[56] Recovery refers to efforts aimed at restoring an area to its prior status, including the reconstruction of damaged structures, including its housing stock, business establishments, public facilities, and the environment. The availability of funds is critical to these efforts; however, FEMA does not conduct an assessment of a jurisdiction's fiscal capacity to fund a recovery effort without federal assistance before determining whether to award federal assistance.

As we previously reported and continue to believe, Total Taxable Resources (TTR), a measure developed by the U.S. Department of the Treasury, provides a more comprehensive measure of a jurisdiction's fiscal capacity than FEMA's current PA indicator. 57 For example, TTR includes much of the business income that does not become part of the income flow to jurisdiction residents, undistributed corporate profits, and rents and interest payments made by businesses to out-of-jurisdiction real estate owners and lenders. In the case of FEMA's PA program, adjustments for TTR in setting the threshold for a disaster declaration could result in a more realistic estimate of a jurisdiction's ability to respond to a disaster. Furthermore, since TTR provides estimates of each jurisdiction's fiscal capacity, adjustments for TTR growth would vary by jurisdiction. FEMA could also use other measures of fiscal capacity, such as state personal income or gross state product, to more accurately determine a jurisdiction's ability to pay for damages to public structures without federal assistance.[58] Table 5 describes three potential approaches to measure a jurisdiction's fiscal capacity. Federal departments and agencies have used some of these approaches to help determine a jurisdiction's fiscal capacity and the extent to which a jurisdiction should be eligible for federal assistance. For example, the Department of Health and Human Services' Substance Abuse and Mental Health Services Administration's block grant program and Community Mental Health Service use TTR. Also, personal income is used by many federal grant programs.

Table 5. Three Potential Approaches to Measure a Jurisdiction's Fiscal Capacity

Potential approach	Description
State Personal Income	Personal income is the income received by all persons from all sources. Personal income is the sum of net earnings by place of residence, property income, and personal current transfer receipts.a
Gross State Product (GSP)	Also known as Gross Domestic Product by state, GSP is the state counterpart of the nation's gross domestic product (GDP), the Bureau of Economic Analysis' featured and most comprehensive measure of U.S. economic activity. GDP by state is derived as the sum of the GDP originating in all the industries in a state.b
Total Taxable Resources	TTR is the unduplicated sum of the income flows produced within a state (GSP) and the income flows received by its residents (personal resident income) that a state can potentially tax. TTR does not consider the actual fiscal choices made by states. It measures all income flows a state can potentially tax.c

Source: GAO analysis of Bureau of Economic Analysis and U.S. Department of Treasury data.

[a] http://www.bea.gov/newsreleases/regional/spi/sqpi_newsrelease.htm

[b] http://www.bea.gov/newsreleases/regional/gdp_state/gsp_newsrelease.htm, and

[c] http://www.treasury.gov/resource-center/economic-policy/taxable-resources/Pages/Total-Taxable-Re sources.aspx.

Without an accurate assessment of a jurisdiction's capabilities to respond to and recover from a disaster without federal assistance, including a jurisdiction's preparedness capabilities and fiscal capacity, FEMA runs the risk of recommending that the President award federal assistance to jurisdictions that have the capability to respond and recover on their own.

Reexamining the basis for the PA indicator and the usefulness of preparedness metrics and jurisdiction fiscal capacity could help FEMA more accurately determine whether a jurisdiction should be eligible for federal assistance. In appendix IV, we provide additional information about the three approaches to measure a jurisdiction's fiscal capacity as well as examples of how these fiscal measures could assist FEMA in more accurately determining whether the magnitude of damage is beyond the capacity of the jurisdiction.

FEMA Does Not Have Specific Criteria to Evaluate Some Cost Share Adjustment Requests and Does Not Track Additional Costs for All Adjustments

According to the Stafford Act, the usual cost share arrangement for disaster declarations calls for the federal government to pay not less than 75 percent of the eligible PA costs of a disaster and nonfederal entities (that is, state and local governments) to pay the remaining 25 percent; at a governor's request, the President can adjust this cost share.[59] FEMA has specific criteria to evaluate a request to adjust the federal share from 75 percent to 90 percent, but does not have specific criteria to evaluate a request to adjust the federal share to 100 percent. Adjusting the federal share to 100 percent is typically done for emergency work such as life- saving activities and debris removal projects through FEMA's PA program. In addition, FEMA does not know the additional costs (that is, the costs of paying an additional 15 or 25 percent) associated with either type of cost share adjustment because the agency does not track these costs.

Governors can request that the President reduce the 25 percent cost share for nonfederal governments to 10 percent or 0 percent. 60 FEMA generally follows the same process to evaluate a request from a governor for a cost share adjustment as it follows to evaluate a request for a disaster declaration, according to FEMA officials. FEMA makes a recommendation to the President as to whether the request for a cost share adjustment should be approved or denied and the President makes the decision.

For the 539 disaster declarations during fiscal years 2004 through 2011, governors requested that the President adjust the usual federal/nonfederal (that is, state and local government) cost share 150 times. As shown in table 6, 109 of the 150 requests, or 73 percent, were approved during this period. However,

23 of the 109 cost share adjustments were required by provisions in law; therefore, FEMA's recommendation was not a factor in whether these cost share adjustment requests were approved or denied.[61] For example, 10 of the 23 cost share adjustments required by law were for Hurricanes Katrina, Wilma, Dennis, and Rita.[62]

Table 6. Number of Cost Share Adjustments Requested and Approved, Fiscal Years 2004 through 2011

Fiscal year	Number requested	Number approved	Percent approved
2004	29	22	76
2005	46	42	91
2006	7	5	71
2007	9	3	33
2008	23	17	74
2009	15	7	47
2010	9	6	67
2011	12	7	58
Total	150	109	73

Source: GAO analysis of FEMA data.

Note: For the number of cost share adjustments requested and approved, we used the fiscal year in which the disaster declaration was approved. Of the 109 cost share adjustments, 23 were required by legislation. Excluding the 23 cost share adjustments, the President approved 86 of the 127 cost share adjustment requests, or 68 percent.

Our analysis shows that 64 of the 109 cost share adjustments during fiscal years 2004 through 2011 were for the following six disasters: 23 for Hurricane Katrina, 11 for Hurricane Rita, 9 for Hurricane Ike, 8 for Hurricane Ivan, 7 for midwest flooding in fiscal year 2008, and 6 for Hurricane Dennis. Furthermore, 34 of the 109 cost share adjustments involved a single adjustment, whereas 23 cost share adjustments involved multiple adjustments. FEMA officials explained this by stating that a 100 percent cost share adjustment could be approved for a 72-hour period and the governor could subsequently request another 100 percent cost share adjustment for another 72-hour period, which the President could approve. For example, the disaster declaration in Louisiana in fiscal year 2005 for Hurricane Katrina had 8 cost share adjustments and the disaster declaration in Mississippi in fiscal year 2005 for Hurricane Katrina had 9 cost share adjustments.

According to FEMA officials, although the process is similar, the agency uses different criteria to evaluate a request from a governor to increase the federal government share for PA up to 90 percent than it does for requests up to 100 percent. Specifically, FEMA may recommend to the President that the federal cost share be increased up to 90 percent when a disaster is so extraordinary that actual federal obligations, excluding FEMA administrative costs, meet or exceed a qualifying threshold. To determine the threshold, the jurisdiction population is multiplied by a per capita amount, which is $135 for calendar year 2012 (or 100 times the 2012 per capita damage indicator of $1.35).[63] Forty-one of the 109 cost share adjustments increased the federal cost share to 90 percent and reduced the nonfederal share to 10 percent. According to FEMA's regulations, if warranted by the needs of the disaster, FEMA may recommend up to 100 percent federal funding for emergency work, such as debris removal and emergency protective measures, for a limited period in the initial days of the disaster irrespective of the per capita amount.[64] Sixty-eight of the 109 cost share adjustments increased the federal cost share to 100 percent.

Unlike its evaluation of a request that the federal share be increased from 75 percent up to 90 percent, FEMA does not use specific criteria to evaluate requests to adjust the federal cost share up to 100 percent. FEMA officials stated that a recommendation to the President for up to a 100 percent cost share adjustment is based on a subjective assessment of the jurisdiction's needs and that it is usually pretty obvious when a jurisdiction needs debris removal and emergency protective measures, although the officials acknowledged that FEMA's recommendation is a judgment call. According to FEMA, it does not use the same criteria to evaluate a request for a 100 percent cost share adjustment as it uses for a 90 percent cost share adjustment because the criteria for the 90 percent adjustment are based on actual federal obligations.

FEMA officials explained that they would not be able to apply those criteria for the 100 percent adjustment in the initial days of a disaster because there would not be much, if any, funding obligated at that point. However, criteria for assessing a request for a 100 percent cost share adjustment for PA (that is, emergency work) do not have to be the same criteria FEMA uses to assess requests for 90 percent cost share adjustments. For example, FEMA's IA grant program uses multiple factors to determine whether to recommend to the President that a jurisdiction be granted IA.[65]

We have previously reported that clear criteria are important for controlling federal costs and helping to ensure consistent and equitable eligibility determinations.[66] For example, if a 100 percent cost share adjustment is approved, the federal government could pay millions of dollars more than it ordinarily would for a single disaster declaration. Furthermore, *Standards for Internal Control in the Federal Government* state that internal control activities help ensure that management's directives are carried out and that actions are taken to address risks.[67]

Moreover, internal control standards state that control activities should be an integral part of an entity's accountability for stewardship of government resources. Without such activities, FEMA is at risk that its recommendations related to 100 percent cost share adjustments may not be justified. Further, relying on professional judgment only, FEMA is at risk of making inconsistent, and potentially inequitable, recommendations to the President about whether to grant 100 percent cost share adjustments.

In addition, FEMA officials stated that they do not know the costs associated with the 109 cost share adjustments because the agency does not track the costs for all cost share adjustments, although on rare occasions, at the request of congressional staff, FEMA officials have identified the costs associated with cost share adjustments, such as those for Hurricane Katrina. The officials stated that they have not routinely tracked the additional costs associated with cost share adjustments because they did not see a need for this information.

According to *Standards for Internal Control in the Federal Government*, program managers need financial data to determine whether they are meeting their goals for accountability for effective and efficient use of resources.[68] Financial information is needed for both external and internal uses, and on a day-to-day basis to make operating decisions, monitor performance, and allocate resources. Pertinent information should be identified, captured, and distributed in a form and time frame that permits people to perform their duties efficiently.

Because FEMA does not track the costs associated with cost share adjustments, FEMA does not know the financial impact of its recommendations to the President on whether to increase the federal cost share for PA. Understanding the financial impact of FEMA's recommendations to the President for cost share adjustments would enable FEMA to make more informed recommendations and estimate the impact of the adjustments on available DRF balances.

COSTS OF PROVIDING DISASTER ASSISTANCE HAVE INCREASED, BUT FEMA IS WORKING TO REDUCE COSTS

FEMA's administrative cost percentages have often surpassed its targets for all sizes of disasters and have doubled in size since fiscal year 1989. FEMA provided guidance for administrative cost targets but does not assess how well the targets were achieved. The agency is working on three short- and long-term initiatives to deliver disaster assistance in a more efficient manner.

Administrative Cost Percentages Often Exceeded FEMA's Targets and Have Doubled since Fiscal Year 1989

Our analysis of the 539 disaster declarations during fiscal years 2004 through 2011 shows that 37 percent of the declarations exceeded administrative cost percentage targets established in guidance prepared by FEMA in 2010.[69] Administrative cost percentages varied widely among disaster declarations that required a similar amount of federal financial assistance, suggesting that certain declarations may have been administered more efficiently than others. In addition, FEMA's average administrative cost percentage for disaster declarations has doubled since fiscal year 1989.[70]

FEMA's administrative costs relate to the delivery of disaster assistance programs, such as the PA or IA programs, and are primarily obligated from the DRF.[71] Examples of administrative costs include the salary and travel costs for the disaster reserve workforce, rent and security expenses associated with JFO facilities, and supplies and information technology support for JFO staff. According to FEMA officials, the agency's administrative costs are primarily due to activities at JFOs; however, administrative costs can also be incurred at FEMA regional offices, headquarters, and other locations.[72]

We analyzed actual administrative costs for disaster declarations that were closed as of April 30, 2012, and, for declarations that were still open as of April 30, 2012, we analyzed actual obligations as of April 30, 2012, plus the amount that FEMA projected to obligate in the future until the declarations are eventually closed. [73] FEMA categorizes disaster declarations using three event levels, essentially small, medium, or large based on the amount of federal funding obligated for the disaster, and has established target ranges for administrative cost percentages for each. Our analysis shows that FEMA frequently exceeded the administrative cost percentage targets established by

FEMA guidance for all three sizes of disaster declarations during fiscal years 2004 through 2011. Specifically:

- For small disaster declarations (total obligations of less than $50 million), the target range for administrative costs is 12 percent to 20 percent; for the 409 small declarations that we analyzed, 4 out of every 10 had administrative costs that exceeded 20 percent.
- For medium disaster declarations (total obligations of $50 million to $500 million), the target range for administrative costs is 9 percent to 15 percent; for the 111 declarations that we analyzed, almost 3 out of every 10 had administrative costs that exceeded 15 percent.
- For large disaster declarations (total obligations greater than $500 million to $5 billion), the target range for administrative costs is 8 percent to 12 percent; for the 19 large declarations that we analyzed, over 4 out of every 10 had administrative costs that exceeded 12 percent.[74]

For small declarations that we analyzed, administrative cost percentages averaged 20 percent and ranged from less than 1 percent to 73 percent. Thus, on average, small disaster declarations were within the upper limit of FEMA's target range. However, 12 small declarations had administrative cost percentages greater than 50 percent, which means that FEMA obligated more for administrative costs than for disaster assistance.[75] For example, if FEMA required $6 million to deliver $4 million in disaster assistance to a jurisdiction, then the related administrative cost percentage would be 60 percent of the total DRF obligations of $10 million. For medium declarations that we analyzed, administrative cost percentages averaged 12 percent and were, therefore, in the middle of the target range. However, administrative cost percentages for medium declarations ranged from less than 1 percent to 55 percent and, for 1 medium declaration, FEMA obligated more for administrative costs than for disaster assistance. For large declarations that we analyzed, administrative cost percentages averaged 13 percent—slightly above the upper limit of the target range— and ranged from 3 percent to 25 percent; therefore, none of the large declarations we analyzed had obligations for administrative costs higher than disaster assistance. FEMA's administrative cost percentages also differed significantly depending on the type of assistance delivered to a jurisdiction. For example, for disaster declarations during fiscal years 2004 through 2011, the average administrative cost percentage for disaster declarations that involved only IA was 34 percent, while the average was less than half of that, at 16 percent, for declarations with only PA. Disaster declarations that included both IA and PA had an average administrative cost percentage of 18 percent.[76]

According to FEMA, incidents of similar size and type have witnessed growing administrative costs since 1989. [77] Our analysis of 1,221 small, medium, and large disaster declarations during fiscal years 1989 through 2011 confirms this increase. As discussed in more detail later in this report, administrative costs have increased dramatically because of a number of factors, including the number of staff deployed to a disaster, which tripled during fiscal years 1989 through 2009. Since fiscal year 1989, the average administrative cost percentage for the 1,221 disaster declarations doubled from 9 percent in the 1989-to-1995 period to 18 percent in the 2004-to-2011 period as shown in table 7.

Table 7. Average Administrative Cost Percentages for 1,221 Disaster Declarations during Fiscal Years 1989 through 2011

Fiscal years that declaration wasapproved	Average administrative cost percentage
1989 to 1995	9
1996 to 2003	14
2004 to 2011	18

Source: GAO analysis of FEMA data.

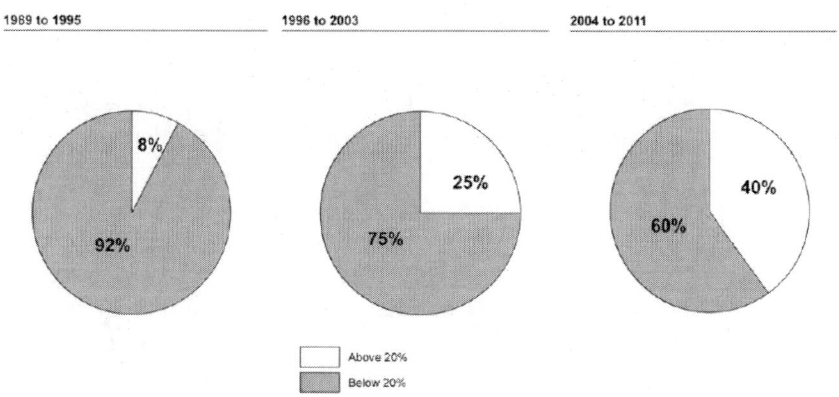

Source: GAO analysis of FEMA data.

Figure 7. Frequency of Declarations above and below FEMA's Target of 20 Percent or Less for Administrative Costs for Small Declarations during Fiscal Years 1989 through 2011.

As shown in figure 7 for the 409 small declarations during fiscal years 2004 through 2011, the frequency of declarations that had administrative cost percentages above FEMA's target of no more than 20 percent is five times what it was during fiscal years 1989 through 1995.[78]

As shown in figure 8 for the 111 medium declarations during fiscal years 2004 through 2011, the frequency of declarations that had administrative cost percentages above FEMA's target range of no more than 15 percent, is almost 10 times what it was during fiscal years 1989 through 1995.

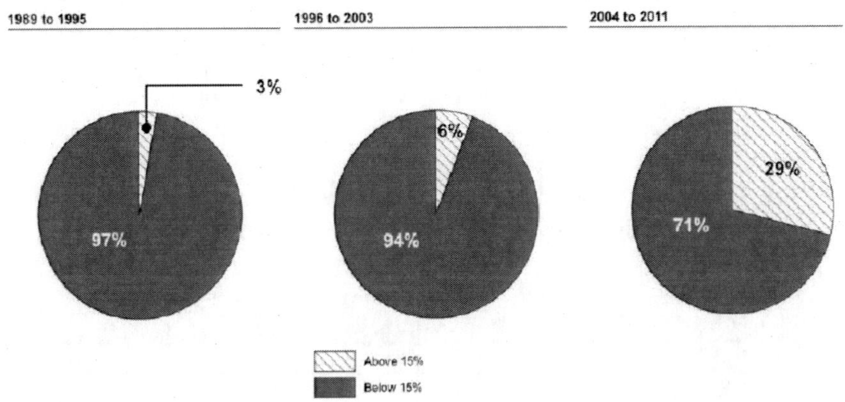

Source: GAO analysis of FEMA data.

Figure 8. Frequency of Declarations above and below FEMA's Target of 15 Percent or Less for Administrative Costs for Medium Declarations during Fiscal Years 1989 through 2011.

As shown in figure 9, for the 19 large declarations during fiscal years 2004 through 2011, the frequency of declarations that had administrative cost percentages above FEMA's target of no more than 12 percent is 42 percent, whereas, during fiscal years 1996 to 2003, 20 percent of the large declarations exceeded FEMA's target and, during fiscal years 1989 through 1995, there were no large declarations that exceeded FEMA's target.

Furthermore, our analysis of administrative cost percentages by type of disaster assistance shows a similar trend of significant increases since fiscal year 1989. As shown in table 8, since fiscal year 1989, administrative cost percentages doubled for declarations with IA only, quadrupled for declarations with PA only, and doubled for declarations with PA and IA.

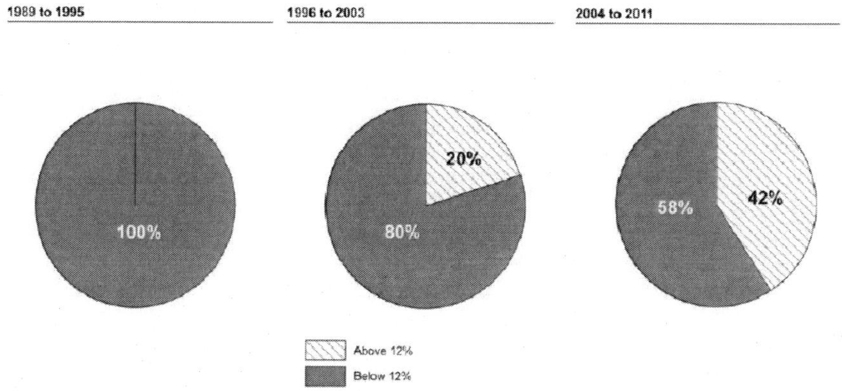

Source: GAO analysis of FEMA data.

Figure 9. Frequency of Declarations above and below FEMA's Target of 12 Percent or Less for Administrative Costs for Large Declarations during Fiscal Years 1989 through 2011.

Table 8. Average Administrative Cost Percentages by Type of Assistance during Fiscal Years 1989 through 2011

Type of assistance	1989-1995	1996-2003	2004-2011
IA	17%	27%	34%
PA	4%	11%	16%
PA, IA	9%	14%	18%

Source: GAO analysis of FEMA data.

FEMA Provided Targets for Administrative Costs in Its Guidance to Staff but Does Not Assess Performance in Achieving the Targets

FEMA officials created a management guide to better manage some of the controllable factors for administrative costs. The guide noted that administrative costs had been steadily rising for 20 years and that little emphasis had been placed on controlling overall costs. The guide set target ranges for administrative costs, but according to FEMA officials, the administrative cost goals are not required to be met. While a number of factors affect FEMA's administrative cost percentages, the agency is working to better manage some of the factors it can control. According to FEMA officials, the amount and type of assistance provided to a jurisdiction are factors that significantly affect administrative cost percentages. Some of the key factors that FEMA can control include the number of staff deployed to a disaster and amount of overtime these staff can work. FEMA officials stated that managing staffing levels according to need throughout the course of a disaster can be a cost-effective alternative to requiring staff to work overtime.

FEMA officials stated that the amount and type of assistance provided to a jurisdiction can affect a disaster declaration's administrative cost percentage. For example, FEMA may be able to achieve economies of scale for relatively large disasters, thereby reducing the related administrative cost percentage. In contrast, relatively small disasters do not benefit from economies of scale and may experience high administrative cost percentages because of the need for a certain number of staff. For instance, according to a FEMA official responsible for controlling disaster-related costs, a typical declaration requires a minimum of 30 to 35 staff to deliver PA and an additional 30 to 35 people to deliver IA.

Therefore, even for a small disaster declaration that includes IA and PA, 60 to 70 staff will be required to administer the assistance. According to FEMA officials, declarations that include IA often experience higher administrative cost percentages. Moreover, FEMA officials stated that the delivery of IA can be more labor intensive than the delivery of PA, as IA often requires one-on-one attention to individuals and families affected by a disaster. Costs associated with a typical JFO, such as rent and employee travel, can be reduced when FEMA utilizes a virtual JFO; however, because of the personal nature of the response required for IA, FEMA is rarely able to utilize a virtual JFO for a declaration with IA according to FEMA officials.[79] Furthermore, while there is no limit on the amount of PA a jurisdiction can receive, the maximum amount an individual/household can receive for IA is $31,400 in fiscal year 2012. As a result of the limit for IA, it is more difficult for FEMA to achieve economies of scale when delivering IA compared with PA.

According to FEMA officials, certain factors that affect administrative costs are managed by the designated FCO.[80] FCOs are primarily responsible for managing JFO operations and planning FEMA's initial disaster response and recovery efforts, including the number of staff needed, and when and how long the staff will be deployed. Salaries, benefits, and travel costs for FEMA staff, as well as costs associated with contractors, account for roughly three-quarters of a disaster declaration's administrative costs.

Also, the average number of FEMA staff deployed to a disaster has increased significantly, from fewer than 500 in fiscal year 1989 to over 1,500 in fiscal year 2009. When we asked FEMA why the staff deployed to a disaster had tripled during this time, FEMA stated that it is extremely difficult to identify any principal factors causing increases in staff and costs because of the complexities associated with the underlying factors, particularly in light of the span of time involved. In addition, for small declarations, a JFO typically closes within 90 days of its opening, and FEMA officials stated that the majority of administrative costs are obligated before a JFO closes. Therefore, planning for and managing the first 90 days of disaster response and recovery can greatly affect the administrative costs for a small declaration.

In November 2010, the FEMA Administrator issued guidance to FCOs on how to better control staffing levels and administrative costs associated with disaster declarations. For example, the document provided guidance on how FCOs can set targets for administrative cost percentages, plan staffing levels, time the deployment of staff, and determine whether to use virtual JFOs. According to FEMA, the guidance is not required to be followed because the agency's intent was to ensure that it was providing guidance to shape how its leaders in the field think about gaining and sustaining efficiencies in operations rather than to lay out a prescriptive formula. In addition, FEMA has a strategic plan that calls for improving the delivery of disaster assistance while minimizing opportunities for waste, fraud, and abuse. To accomplish this, FEMA plans to collaboratively adopt and communicate policies and practices that successfully safeguard against waste, fraud, and abuse. The plan states that, to justify FEMA funding to all of its stakeholders, including taxpayers, there must be a focus on identifying outcomes and measuring performance.

Furthermore, according to the plan, FEMA's ability to analyze and evaluate the results of its plans, programs, and organizational initiatives is key to managing its strategic and long-range organizational goals. However, neither the guidance to FCOs nor the strategic plan sets goals for administrative cost percentages that are required to be met or tracked for performance.

While we recognize complexities exist that can affect administrative costs for disasters, FEMA's implementation of goals to control administrative costs and monitoring of performance in achieving these goals would provide valuable information in assessing the agency's performance in this area and help to provide reasonable assurance that FEMA is delivering disaster assistance in an efficient manner.

FEMA Is Working to Increase Efficiencies in Delivering Disaster Assistance

FEMA officials are considering other short- and long-term actions to increase efficiencies in the delivery of disaster assistance. According to FEMA officials, on the basis of a recent review of the PA program, the agency created a plan with three initiatives focused on delivering assistance to jurisdictions in a more efficient manner. The first initiative under this plan attempts to promote consistency across FEMA's 10 regional offices. For example, FEMA plans to issue a *Public Assistance Program Field Operations Pocket Guide* to provide consistent guidance and approaches to regional staff to help ensure that PA policies are being consistently followed across FEMA regions. FEMA provided the *PA Pocket Guide* to FEMA regional staff and state, local, and tribal officials who attended FEMA's PA Conference and State Workshop in April 2012 to solicit feedback.

The second initiative attempts to reduce the problems associated with debris removal. For example, the agency hopes to reduce its reliance on the Army Corps of Engineers for debris removal, which FEMA officials said can be costlier than hiring local contractors. According to FEMA officials, the agency plans to implement this initiative during the next 2 to 3 years and it may require regulatory changes.

The third initiative addresses inefficiencies within the permanent work portion of PA and includes the consideration of a new model to deliver this assistance. In October 2011, the FEMA Administrator testified that the agency is considering a new, two-step process wherein FEMA and the state would agree on the estimated damages for a large-scale project and FEMA would make a one-time payment to the state.[81] The Administrator stated that the agency currently uses a reimbursement program, which requires a high degree of FEMA oversight and administrative costs.[82] While FEMA officials told us that the agency is in the conceptual stages of developing the new model, officials have identified some potential benefits. For example, the new model could lower FEMA's administrative costs and the time needed to close a disaster declaration. In addition, the new model could allow states to rebuild as they see fit rather than requiring states to replace damaged structures to the status that existed prior to the disaster, which is FEMA's current practice. According to FEMA officials, implementation of this model could require regulatory and legislative changes and is, therefore, years away.

CONCLUSION

Disaster declarations have increased over recent decades, and FEMA has obligated over $80 billion in federal assistance for disasters declared during fiscal years 2004 through 2011, highlighting the importance of FEMA's assessment of jurisdictions' capabilities to respond and recover without federal assistance. The PA per capita indicator is artificially low because it does not reflect the rise in per capita personal income since 1986 or 13 years of inflation from 1986, when the indicator was set at $1.00 and adopted for use, to 1999. By primarily relying on an artificially low indicator, FEMA's recommendations to the President are based on damage estimates and do not comprehensively assess a jurisdiction's capability to respond to and recover from a disaster on its own. For example, on the basis of FEMA's actual and estimated disaster assistance obligations, more than one-third of the 539 major disasters declared during fiscal years 2004 through 2011 are expected to have total DRF obligations of less than $10 million, and more than 60 percent are expected to have total obligations of less than $25 million. Therefore, many of these declarations were for relatively small disasters. At a minimum, adjusting the existing PA per capita indicator fully for changes in per capita income or inflation could ensure that the per capita indicator more accurately reflects changes in U.S. economic conditions since 1986, when the indicator was adopted. Making the appropriate inflation adjustment to the indicator would raise it from $1.35 to $2.07. A change of this size in 1 year could present challenges for jurisdictions, which could find that disasters with PA damage estimates that would now qualify for PA would no longer qualify. Thus, phasing in the adjustment over several years could provide jurisdictions time to take actions, such as increasing any rainy day funds, to adjust to the effects of higher qualifying indicators.

A more comprehensive approach to determine a jurisdiction's capabilities to respond to a disaster would be to replace or supplement the current indicator with more complete data on a jurisdiction's fiscal resources, such as TTR, and would be informed by data on a jurisdiction's response and recovery assets and capabilities. Because FEMA's current approach of comparing the amount of disaster damage with the PA per capita indicator does not accurately reflect whether a jurisdiction has the capabilities to respond to and recover from a disaster without federal assistance, developing a methodology that provides a more comprehensive assessment of jurisdictions' response and recovery capabilities, including a jurisdiction's fiscal capacity, could provide FEMA with data that are more specific to the jurisdiction requesting assistance. For example, developing preparedness metrics in response to the Post-Katrina Act and Presidential Policy Directive-8 could provide FEMA with readily available information on jurisdictions' response and recovery capabilities. Without an accurate assessment of jurisdictions' capabilities to respond to and recover from a disaster, FEMA runs the risk of recommending to the President that federal disaster assistance be awarded without considering a jurisdiction's response and recovery capabilities or its fiscal capacity. As we recommended in 2001, we continue to believe that FEMA should develop more objective and specific criteria to assess the capabilities of jurisdictions to respond to a disaster. Given the legislative and policy changes over the past decade, we believe that including fiscal and nonfiscal capabilities, including available preparedness metrics in its assessment, would allow FEMA to make more informed recommendations to the President when determining a jurisdiction's capacity to respond without federal assistance.

Making informed recommendations to the President about whether cost share adjustments should be granted is important for FEMA and the requesting jurisdictions because every cost share adjustment has financial implications for both entities. A specific set of criteria or factors to use when considering requests for 100 percent cost share adjustments would provide FEMA a decision-making framework and enable more consistent and objectively based recommendations to the President. Also, when FEMA recommends that a cost share adjustment be approved and the President approves it, the federal government assumes the financial burden of paying 15 percent or 25 percent more in PA, which could total millions of dollars. Tracking the additional costs to the federal government because of cost share adjustments would allow FEMA to better understand the financial implications of its recommendations to the President.

FEMA's average administrative costs as a percentage of total DRF disaster assistance obligations have risen for disasters of all sizes. The agency recognized that delivering assistance in an efficient manner is important and published guidance to be used throughout the agency to help rein in administrative costs. However, FEMA has not implemented the goals and does not track performance against them. Over time, reducing administrative costs could save billions of dollars—dollars that could be used to fund temporary housing, infrastructure repairs, and other disaster assistance. Therefore, incentivizing good management over administrative costs by adopting administrative cost percentage goals and measuring performance against these goals would help provide FEMA with additional assurance that it is doing its utmost to deliver disaster assistance in an efficient manner.

RECOMMENDATIONS FOR EXECUTIVE ACTION

To increase the efficiency and effectiveness of the process for disaster declarations, we recommend that the FEMA Administrator take the following four actions:

1) Develop and implement a methodology that provides a more comprehensive assessment of a jurisdiction's capability to respond to and recover from a disaster without federal assistance. This should include one or more measures of a jurisdiction's fiscal capacity, such as TTR, and consideration of the jurisdiction's response and recovery capabilities. If FEMA continues to use the PA per capita indicator to assist in identifying a jurisdiction's capabilities to respond to and recover from a disaster, it should adjust the indicator to accurately reflect the annual changes in the U.S. economy since 1986, when the current indicator was first adopted for use. In addition, implementing the adjustment by raising the indicator in steps over several years would give jurisdictions more time to plan for and adjust to the change.
2) Develop and implement specific criteria or factors to use when evaluating requests for cost share adjustments that would result in the federal government paying up to 100 percent of disaster declaration costs.
3) Annually track and monitor the additional costs borne by the federal government for the cost share adjustments.
4) Implement goals for administrative cost percentages and monitor performance to achieve these goals.

AGENCY COMMENTS AND OUR EVALUATION

We provided a draft of this report to DHS for comment. We received written comments from DHS on the draft report, which are summarized below. DHS concurred with three recommendations and partially concurred with the fourth recommendation.

Regarding the first recommendation, that FEMA develop and implement a methodology that provides a more comprehensive assessment of a jurisdiction's capability to respond to and recover from a disaster without federal assistance, DHS concurred. DHS stated that a review of the criteria used to determine a state's response, recovery, and fiscal capabilities is warranted and that such a review would include the need to update the per capita indicator as well as a review of alternative metrics. DHS stated that any changes would need to be made through the notice and comment rulemaking process and that, if changes are made to the per capita indicator, FEMA's Office of Response and Recovery will review the feasibility of phasing them in over time. However, the extent to which the planned actions will fully address the intent of this recommendation will not be known until the agency completes its review and implements a methodology that provides a more comprehensive assessment of a jurisdiction's capability to respond and, if the per capita indicator continues to be used, adjusts the per capita indicator to accurately reflect annual changes in the U.S. economy since 1986. We will continue to monitor DHS's efforts.

Regarding the second recommendation, that FEMA develop and implement specific criteria or factors to use when evaluating requests for cost share adjustments that would result in the federal government paying up to 100 percent of disaster declaration costs, DHS concurred with the recommendation and stated that FEMA's Office of Response and Recovery will review specific cost share factors or criteria and develop guidelines to support decision making. These actions, if implemented effectively, should address the intent of the recommendation.

DHS concurred with the third recommendation, to track and monitor the additional costs associated with cost share adjustments, and stated that FEMA's Office of Response and Recovery will be responsible for tracking these costs on an annual basis. DHS stated that such actions would provide valuable information for budgetary purposes and for decision makers who consider requests for cost share adjustments. We agree. Thus, these actions, if implemented effectively, should address the intent of the recommendation.

DHS partially concurred with the fourth recommendation, to implement goals for administrative cost percentages and monitor performance to achieve these goals. Specifically, DHS stated that it agrees that setting goals and monitoring performance for achieving these goals is a good practice in any program and can help ensure more effective and efficient operations. However, DHS stated that it plans to conduct a review to better understand and describe its current measures. DHS stated that a number of factors affect administrative costs, which can present difficulties when trying to implement a simple measure of percentage of administrative costs to total costs. For example, DHS noted that the types of assistance provided and the location of the JFO would affect the percentage of administrative costs. DHS also stated that establishing meaningful administrative cost percentage goals will be challenging because of the many factors involved and that a suite of measures to track administrative cost percentages could help ensure more effective and efficient operations. Thus, DHS is pursuing development of such a suite of measures. We agree that a number of factors affect the percentage of administrative costs and that establishing meaningful administrative cost percentage goals can be challenging. In developing a suite of measures, it is important that FEMA's leadership be able to use them to effectively monitor a disaster declaration's overall administrative costs in addition to the factors that affect administrative costs. If these measures allow FEMA to monitor overall administrative costs as well as the factors that affect such costs, then development and implementation of such measures should meet the intent of the recommendation.

DHS also provided technical comments, which we incorporated, as appropriate.

William O. Jenkins, Jr.
Director
Homeland Security and Justice

APPENDIX I: GAO ANALYSES OF THE NUMBER OF DISASTER DECLARATIONS AND AMOUNT OF OBLIGATIONS

Disaster declarations can take a decade or more to close because of a number of factors, including the Federal Emergency Management Agency's (FEMA) reimbursement process for Public Assistance (PA) infrastructure projects, which can take a long time to finish. In addition, some projects are delayed because of disagreements, and sometimes litigation, over the appropriate amount that should be obligated, according to FEMA officials. As shown below in table 9, the oldest open disaster dates back to 1992, making it 20 years old, and only 54 percent of disaster declarations from fiscal year 2001 were closed as of January 31, 2012. All disaster declarations prior to fiscal year 1992 are closed. Table 9 shows the number of major disaster declarations by fiscal year that were open as of January 31, 2012, and the percentage of declarations that have been closed for each year since fiscal year 1992.[1]

Table 9. Status of Disaster Declarations during Fiscal Years 1992 through 2011, as of January 31, 2012

	Number of disaster declarations			
Fiscal year	Declared	Open	Closed	Percentage closed
1992	46	1	45	98
1993	39	1	38	97
1994	36	2	34	94
1995	29	4	25	86
1996	72	2	70	97
1997	49	4	45	92
1998	61	8	53	87
1999	53	10	43	81
2000	40	9	31	78
2001	50	23	27	54
2002	42	16	26	62
2003	62	29	33	53
2004	65	46	19	29
2005	45	31	14	31
2006	53	46	7	13
2007	68	65	3	4
2008	68	68	0	0
2009	63	63	0	0
2010	79	79	0	0
2011	98	98	0	0

Source: GAO analysis of FEMA data.

Fifty-nine jurisdictions received major disaster declarations during fiscal years 1953 through 2011. Texas had the most, with 86 declarations, while Palau had 1. Wyoming, Utah, and Rhode Island had the fewest declarations for a state, each with 9 declarations. Table 10 identifies the number of disaster declarations for all jurisdictions during fiscal years 1953 through 2011.

The number of major disaster declarations and total obligations varied among FEMA regions during fiscal years 2004 through 2011. For example, FEMA Region X had 32 declarations, while FEMA Region IV had 87. In addition, obligations for FEMA Region VI during this time, which was affected by Hurricane Katrina, reached nearly $40 billion, while FEMA Region X had obligations of $647 million. See figure 10 for more information.

Table 10. Number of Disaster Declarations during Fiscal Years 1953 through 2011 by Jurisdiction

Jurisdiction	Number of declarations
Texas	86
California	78
Oklahoma	70
New York	65
Florida	63
Louisiana	57
Alabama	55
Kentucky	55
Arkansas	53
Missouri	53
Illinois	51
Mississippi	50
Tennessee	50
Iowa	48
Minnesota	48
Kansas	47
Nebraska	47
Pennsylvania	46
West Virginia	46
Ohio	45
Washington	44
Virginia	43
North Dakota	42
North Carolina	40
Indiana	39
Maine	39
South Dakota	39
Georgia	36
Alaska	35
Wisconsin	35
Vermont	32
New Jersey	31
Oregon	28
New Hampshire	27
Massachusetts	26
Hawaii	25
Michigan	25
Puerto Rico	25
Federated States of Micronesia	24

Table 10. (Continued)

Jurisdiction	Number of declarations
Arizona	23
Idaho	23
New Mexico	23
Maryland	21
Montana	20
Nevada	17
Virgin Islands	17
Colorado	16
Connecticut	16
South Carolina	15
Delaware	14
Northern Mariana Islands	14
Guam	12
American Samoa	11
District of Columbia	10
Rhode Island	9
Utah	9
Wyoming	9
Marshall Islands	7
Palau[a]	1

Source: GAO analysis of FEMA data.

[a] We did not include Palau in our analysis of disaster declarations during fiscal years 2004 through 2011 because Palau is a Freely Associated State that, under the terms of its Compact of Free Association with the United States, ceased to receive disaster assistance from the U.S. government. Prior to entering the compact, Palau received one disaster declaration as a territory.

FEMA classifies major disaster declarations by incident type, and these types include floods, tornadoes, and hurricanes, among other types of disasters—both natural and man-made. As shown in table 11, the most frequent type of incident was, according to FEMA data, severe storms, which accounted for 71 percent of the declarations during fiscal years 2004 through 2011.

FEMA obligates funds from the Disaster Relief Fund (DRF) to help jurisdictions respond to and recover from declared disasters. FEMA classifies these funds into five categories: PA, Individual Assistance, Hazard Mitigation, Mission Assignments, and Administration. Table 12 shows the obligations for each category by jurisdiction.

Table 11. Disaster Declarations by Incident Type during Fiscal Years 2004 through 2011

Incident type[a]	Number of declarations	Percentage of total
Severe storm	385	71
Hurricane	37	7
Flood	34	6
Fire	21	4
Snow	20	4
Severe ice storm	12	2
Tornado	8	1
Typhoon	7	1
Other	5	1
Earthquake	4	1
Tsunami	3	1
Freezing	2	Less than 0.5
Dam/levee break	1	Less than 0.5
Total	539	100

Source: GAO analysis of FEMA data.

[a]We are reporting the incident type for each disaster declaration as assigned by FEMA.

Table 12. Obligations for Disaster Declarations during Fiscal Years 2004 through 2011, by Jurisdiction, as of January 31, 2012

Jurisdiction	Disaster assistance						
	Public Assistance	Individual Assistance	Hazard Mitigation	Mission Assignment	Total assistance	Administration	Grand total
Alabama	$656	$903	$99	$383	$2,041	$414	**$2,456**
(including Katrina)							
Alabama	536	297	53	320	1,206	228	**1,433**
(excluding							
Katrina)							
Alaska	81	5	7	1	95	25	**120**
American Samoa	74	66	4	9	153	64	**217**
Arizona	43	0	2	0	44	9	**53**
Arkansas	372	53	50	3	477	80	**557**
California	1,057	113	76	4	1,251	117	**1,368**
Colorado	0	1	a	a	1	2	**3**
Connecticut	38	18	2	a	58	28	**85**
Delaware	12	0	1	0	12	2	**14**
District of Columbia	12	0	0	0	12	1	**13**
Federated States of Micronesia	16	8	1	a	25	8	**33**
Florida (including Katrina)	4,904	2,201	498	847	8,449	1,128	**9,578**
Florida (excluding Katrina)	4,687	2,201	490	847	8,225	1,105	**9,331**
Georgia	190	80	57	1	329	94	**422**
Guam	1	0	0	0	1	1	**2**
Hawaii	56	16	4	a	76	17	**93**
Idaho	12	0	a	0	12	4	**17**

Jurisdiction	Disaster assistance						
	Public Assistance	Individual Assistance	Hazard Mitigation	Mission Assignment	Total assistance	Administration	Grand total
Illinois	251	481	23	3	757	98	**855**
Indiana	213	140	40	3	396	66	**463**
Iowa	1,626	227	296	10	2,158	169	**2,327**
Kansas	973	89	94	4	1,160	71	**1,231**
Kentucky	375	102	45	2	524	108	**632**
Louisiana (including Katrina)	13,682	11,353	981	2,742	28,758	3,579	**32,337**
Louisiana (excluding Katrina)	1,770	991	172	328	3,261	587	**3,848**
Maine	74	4	6	a	84	21	**$105**
Maryland	96	0	2	0	98	6	**104**
Massachusetts	173	99	9	1	282	58	**340**
Michigan	15	45	4	0	64	8	**72**
Minnesota	168	38	13	3	222	41	**264**
Mississippi (including Katrina)	3,561	3,111	296	1,207	8,174	2,072	**10,246**
Mississippi (excluding Katrina)	107	54	6	13	180	56	**236**
Missouri	592	112	66	152	923	137	**1,060**
Montana	32	7	1	0	40	17	**57**
Nebraska	292	8	29	a	329	37	**366**
Nevada	21	2	2	a	25	5	**30**
New Hampshire	91	15	12	a	119	28	**147**
New Jersey	193	232	37	2	464	71	**535**
New Mexico	57	8	1	0	66	17	**82**
New York	839	266	40	22	1,168	152	**1,320**
North Carolina	152	102	5	1	259	70	**329**
North Dakota	437	209	21	42	710	146	**856**

Table 12. (Continued)

Jurisdiction	Disaster assistance						
	Public Assistance	Individual Assistance	Hazard Mitigation	Mission Assignment	Total assistance	Administration	Grand total
Northern Mariana Islands	10	14	2	0	26	6	**32**
Ohio	238	92	29	1	359	45	**404**
Oklahoma	463	37	38	1	539	82	**622**
Oregon	106	9	19	0	134	24	**159**
Pennsylvania	276	298	25	10	610	85	**695**
Puerto Rico	152	485	17	9	664	99	**763**
Rhode Island	27	42	4	a	73	18	**91**
South Carolina	43	8	3	0	54	7	**61**
South Dakota	200	13	8	1	222	52	**273**
Tennessee	332	203	56	2	593	118	**711**
Texas	2,795	1,843	395	280	5,313	1,056	**6,369**
Utah	28	0	1	0	28	7	**36**
Vermont	51	28	3	1	82	36	**118**
Virgin Islands	18	0	1	0	19	8	**27**
Virginia	144	14	6	a	164	22	**187**
Washington	209	43	41	a	293	58	**352**
West Virginia	108	117	19	12	256	58	**314**
Wisconsin	113	138	30	a	282	50	**332**
Wyoming	8	3	a	0	11	3	**15**
Total	**$36,759**	**$23,500**	**$3,522**	**$5,761**	**$69,542**	**$10,806**	**$80,349**

Source: GAO analysis of FEMA data.

Note: Total obligations include obligations for Hurricane Katrina and do not factor in jurisdictions' obligations that exclude Hurricane Katrina.

[a]Obligations less than $0.5 million.

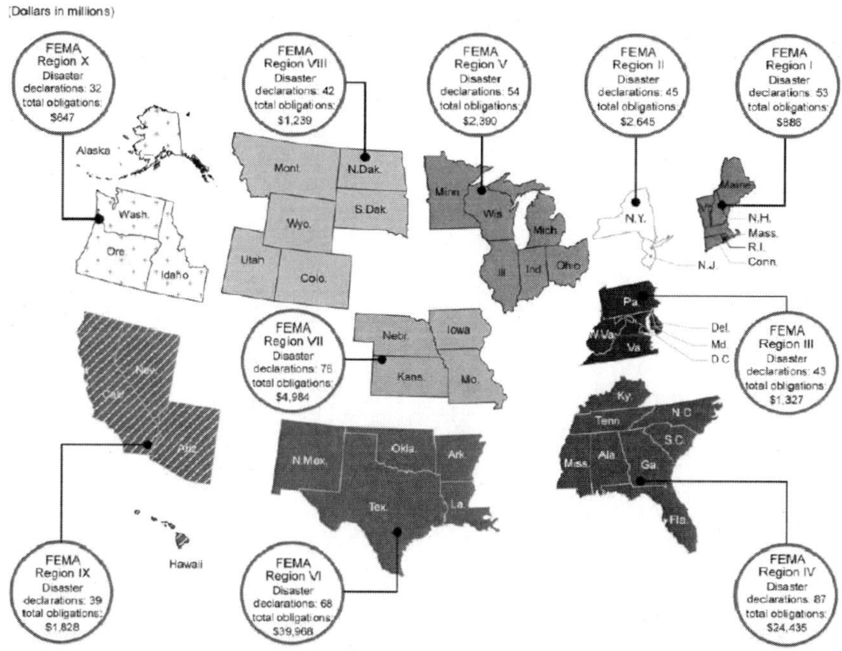

Source: GAO analysis of FEMA data, and Map Resources (map).

Note: American Samoa, Federated States of Micronesia, Guam, Marshall Islands, U.S. Virgin Islands, Commonwealth of the Northern Mariana Islands, and Commonwealth of Puerto Rico are not visually represented in the figure, but the number of disaster declarations and obligations due to these jurisdictions were included in the figure under Regions II and IX. Specifically, the U.S. Virgin Islands and Commonwealth of Puerto Rico are served by FEMA Region II, and America Samoa, Federated States of Micronesia, Guam, Marshall Islands, and Commonwealth of the Northern Mariana Islands are served by FEMA Region IX.

Figure 10. Number of Disaster Declarations and Amount of Obligations by FEMA Region during Fiscal Years 2004 through 2011.

Obligations on a per person basis varied for disasters declared during fiscal years 2004 through 2011. For example, including Hurricane Katrina, Louisiana had the highest per capita obligations at $7,236, but excluding obligations for Hurricane Katrina, American Samoa had the highest obligations at $3,795 per person. For the lowest obligations per person, Colorado had 81 cents and the Marshall Islands had zero. See tables 13 and 14 for obligations on a per person basis for all 58 jurisdictions when including and excluding obligations for Hurricane Katrina, respectively.

Table 13. Obligations Per Person Including Hurricane Katrina for Disaster Declarations during Fiscal Years 2004 through 2011, by Jurisdiction

Rank order	Jurisdiction	Obligated per person as of January 31, 2012
1.	Louisiana	$7,236
2.	American Samoa	3,795
3.	Mississippi	3,602
4.	North Dakota	1,332
5.	Iowa	795
6.	Florida	599
7.	Alabama	552
8.	Northern Mariana Islands	460
9.	Kansas	458
10.	South Dakota	362
11.	Federated States of Micronesia	307
12.	Texas	305
13.	Virgin Islands	245
14.	Nebraska	214
15.	Arkansas	208
16.	Puerto Rico	200
17.	Vermont	194
18.	Alaska	191
19.	Missouri	189
20.	Oklahoma	180
21.	West Virginia	174
22.	Kentucky	156
23.	Tennessee	125
24.	New Hampshire	119
25.	Rhode Island	87
26.	Maine	82
27.	Hawaii	77
28.	Indiana	76
29.	New York	70
30.	Illinois	69
31.	New Jersey	64
32.	Montana	63
33.	Wisconsin	62
34.	Washington	60
35.	Pennsylvania	57
36.	Minnesota	54
37.	Massachusetts	54

Rank order	Jurisdiction	Obligated per person as of January 31, 2012
38.	Georgia	52
39.	Oregon	46
40.	New Mexico	45
41.	North Carolina	41
42.	California	40
43.	Ohio	36
44.	Wyoming	29
45.	Virginia	26
46.	Connecticut	25
47.	District of Columbia	22
48.	Maryland	20
49.	Delaware	18
50.	Utah	16
51.	South Carolina	15
52.	Nevada	15
53.	Idaho	13
54.	Guam	12
55.	Arizona	10
56.	Michigan	7
57.	Colorado	0.81
58.	Marshall Islands	0
Total		$23,362

Source: GAO analysis of FEMA data.

Table 14. Obligations Per Person Excluding Hurricane Katrina for Disaster Declarations during Fiscal Years 2004 through 2011, by Jurisdiction

Rank order	Jurisdiction	Obligated per person as of January 31, 2012
1.	American Samoa	$3,795
2.	North Dakota	1,332
3.	Louisiana	861
4.	Iowa	795
5.	Florida	584
6.	Northern Mariana Islands	460
7.	Kansas	458
8.	South Dakota	362
9.	Alabama	322
10.	Federated States of Micronesia	307
11.	Texas	305

Table 14. (Continued)

Rank order	Jurisdiction	Obligated per person as of January 31, 2012
12.	Virgin Islands	245
13.	Nebraska	214
14.	Arkansas	208
15.	Puerto Rico	200
16.	Vermont	194
17.	Alaska	191
18.	Missouri	189
19.	Oklahoma	180
20.	West Virginia	174
21.	Kentucky	156
22.	Tennessee	125
23.	New Hampshire	119
24.	Rhode Island	87
25.	Mississippi	83
26.	Maine	82
27.	Hawaii	77
28.	Indiana	76
29.	New York	70
30.	Illinois	69
31.	New Jersey	64
32.	Montana	63
33.	Wisconsin	62
34.	Washington	60
35.	Pennsylvania	57
36.	Minnesota	54
37.	Massachusetts	54
38.	Georgia	52
39.	Oregon	46
40.	New Mexico	45
41.	North Carolina	41
42.	California	40
43.	Ohio	36
44.	Wyoming	29
45.	Virginia	26
46.	Connecticut	25
47.	District of Columbia	22
48.	Maryland	20
49.	Delaware	18
50.	Utah	16

Rank order	Jurisdiction	Obligated per person as of January 31, 2012
51.	South Carolina	15
52.	Nevada	15
53.	Idaho	13
54.	Guam	12
55.	Arizona	10
56.	Michigan	7
57.	Colorado	0.81
58.	Marshall Islands	0
Total		$13,222

Source: GAO analysis of FEMA data.

APPENDIX II: OBJECTIVES, SCOPE, AND METHODOLOGY

This report addresses the following questions: (1) For each fiscal year from 2004 through 2011, how many disaster declaration requests did FEMA receive, how many were approved, for which types of disasters, and how much were the associated obligations from the DRF? (2) What criteria has FEMA used to recommend to the President that a disaster declaration is warranted for PA, and to what extent does FEMA assess whether an effective response to a disaster is beyond the capabilities of state and local governments? (3) How does FEMA determine whether a cost share adjustment recommendation for PA is warranted and how much additional federal assistance did jurisdictions receive during fiscal years 2004 through 2011 because of cost share adjustments? (4) What were FEMA's administrative cost percentages for disaster declarations during fiscal years 2004 through 2011, how have they changed over time, and what actions is FEMA taking, if any, to reduce the costs of delivering disaster assistance funds?

To determine how many disaster declaration requests FEMA received, how many were approved, for which types of disasters, and how much the associated obligations were from the DRF, we obtained data for each disaster declaration approved during fiscal years 2004 through 2011.[1] We focused on this time frame because it contains the most current data for disaster declarations. It also comprises the time period after FEMA was merged into the newly created DHS, on March 1, 2003, and predates Hurricane Katrina in 2005. We focused primarily on fiscal years 2004 through 2011; however, to provide historical context and to compare results across similar periods, we also reviewed obligations data from fiscal years 1989 through 2011. In addition, to provide further historical perspective, we include information on the number of disaster declarations by jurisdiction from the first presidential disaster declaration in fiscal year 1953 through fiscal year 2011 in appendix I. FEMA provided data to us from its National Emergency Management Information System (NEMIS) and Integrated Financial Management Information System (IFMIS). To determine whether the data were reliable, we reviewed the data that FEMA officials provided and discussed data quality control procedures to ensure the integrity of the data with them. We determined that the data we used from these systems were sufficiently reliable for the purposes of this report.

To determine the criteria that FEMA used to recommend to the President that a disaster declaration was warranted for PA, and to what extent FEMA assessed whether an effective response to a disaster was beyond the capabilities of jurisdictions, such as state and local governments, we examined FEMA policies, regulations, and other documents related to the disaster declarations process. To determine the probability of a disaster declaration request being approved for PA if the Preliminary Damage Assessments (PDA) met or exceeded the PA per capita indicator, we obtained and analyzed data on FEMA's PDAs from fiscal years 2008 through 2011. For this analysis, we used 4 years of data (fiscal years 2008 through 2011) instead of 8 years of data (fiscal years 2004 through 2011) that we used for other analyses because FEMA did not have data for fiscal years 2004 through 2007 in an electronic format. We believe that our analysis of 4 years of data is sufficient for purposes of this report. Specifically, we analyzed 246 disaster declarations during fiscal years 2008 through 2011, and excluded 293 declarations during fiscal years 2004 through 2007 because FEMA had readily available data only for PDAs for fiscal years 2008 through 2011. For each of the 246 disaster declarations, we reviewed the PDAs to determine whether a state requested PA, whether the President approved it, and the extent to which the PA damage estimate exceeded the PA per capita indicator.

In addition, we conducted an analysis to determine whether disaster declarations from 2004 through 2011 would have met the PA per capita indicator if adjusted for the change in per capita personal income since 1986. Our analysis included FEMA's projected obligations as of April 30, 2012, for only those 508 disaster declarations that had received PA and had been declared during fiscal years 2004 through 2011. We did not analyze the 31 disaster declarations that only received IA. We analyzed obligations instead of PDA damage estimates for PA because FEMA officials stated that estimating the damage from a disaster is sometimes stopped when the estimate equals or exceeds the PA per capita indicator. Therefore, we concluded that conducting the analysis using projected obligations would be more accurate than using incomplete PDA damage estimates for PA. In addition, we separately analyzed actual obligations for 144 closed disaster declarations because closed declarations would be either complete or very close to being complete. To determine whether the data were reliable, we reviewed the data that FEMA officials provided and discussed data quality control procedures to ensure the integrity of the data with them. We determined that the data we used from PDAs were sufficiently reliable for the purposes of this report.

To determine how FEMA evaluated whether a cost share adjustment recommendation was warranted and how much additional federal assistance states received during fiscal years 2004 through 2011 because of the adjustments, we obtained and reviewed relevant laws, regulations, and policies. We also obtained and analyzed the cost share adjustments and types requested, approved, and denied during fiscal years 2004 through 2011. In addition, we interviewed FEMA officials who process cost share adjustment requests and participate in making recommendations to the President as to whether the requests should be approved or denied. We also reviewed internal control standards for the federal government related to ensuring management directives are carried out and that actions are taken to address risks.[2] To determine whether the data were reliable, we reviewed the data that FEMA officials provided and discussed data quality control procedures to ensure the integrity of the data with them. We determined that the cost share adjustment data were sufficiently reliable for the purposes of this report.

To determine FEMA's administrative cost percentages for disaster declarations, we obtained DRF actual obligations, projected obligations, and related data for all 1,221 disaster declarations from fiscal years 1989 through 2011. While the focus of our objective was fiscal years 2004 through 2011, we obtained obligations data back to fiscal year 1989 to assess potential trends over time because FEMA only maintains obligations data since then. To assess FEMA's current practices, we compared FEMA's administrative cost percentages for disaster declarations during fiscal years 2004 through 2011 with FEMA's target ranges for administrative cost percentages. To identify potential trends over time, we compared FEMA's administrative cost percentages during fiscal years 1989 through 2003 with FEMA's administrative cost percentages during fiscal years 2004 through 2011 and with FEMA's target ranges. According to FEMA officials, administrative costs are typically higher in the early months of a declaration, typically decreasing as the declaration matures (that is, as labor-intensive response activities decline). In order to ensure the results of both analyses were not skewed by declarations that had not yet matured and whose administrative costs were high, we analyzed actual administrative costs for disaster declarations that were closed as of April 30, 2012. For declarations that were still open as of April 30, 2012, we analyzed actual obligations as of April 30, 2012, plus the amount that FEMA projected to obligate in the future until the declarations are eventually closed.

To determine whether the data were reliable, we reviewed the data that FEMA officials provided and discussed data quality control procedures to ensure the integrity of the data with them. We determined that the DRF data were sufficiently reliable for the purposes of this report. To determine what actions FEMA is taking, if any, to reduce the costs of delivering disaster assistance, we interviewed FEMA officials and reviewed relevant policies, documents, and briefings.

In addition to conducting interviews with officials in FEMA headquarters for all four objectives, we conducted site visits to two FEMA regions—Regions IV and VI, which had the highest total obligations during fiscal years 2004 through 2011. The regional administrative offices were located in Atlanta, Georgia, and Denton, Texas, respectively. At each region, we interviewed the Regional or Deputy Administrator and various other personnel.

In addition, we visited the emergency management agencies for Georgia and Oklahoma—one state within each of the two FEMA regions. We selected the two state emergency management agencies—Georgia and Oklahoma—based on their respective proximity to FEMA's regional offices, their high level of experience with disasters, and their availability for a visit during September 2011. We wanted to avoid states that were actively responding to a disaster during that time. While the information we obtained on these site visits is not generalizable, the visits provided important insights into the disaster declaration process.

We conducted this performance audit from July 2011 through September 2012 in accordance with generally accepted government auditing standards. Those standards require that we plan and perform the audit to obtain sufficient, appropriate evidence to provide a reasonable basis for our findings and conclusions based on our audit objectives. We believe that the evidence obtained provides a reasonable basis for our findings and conclusions based on our audit objectives.

APPENDIX III: TOTAL PUBLIC ASSISTANCE INDICATOR AMOUNT FOR FISCAL YEAR 2012, BY JURISDICTION

FEMA uses a PA per capita indicator to help determine a jurisdiction's need for federal assistance in the wake of a disaster. Table 15 shows how the indicator is calculated for each jurisdiction. FEMA multiplies the 2010 population for each jurisdiction by the PA per capita indicator for the fiscal year in which the disaster occurs. In fiscal year 2012, the PA per capita indicator is $1.35. The results of these calculations are the total indicator amounts in table 15. If the PA damage estimate exceeds the total indicator amount, a jurisdiction is likely to receive a major disaster declaration.

Table 15. 2010 Population and Total PA Per Capita Indicator Amount for Fiscal Year 2012, by Jurisdiction

Jurisdiction	2010 population	Total indicator amount (based on $1.35 per capita)a
California	37,253,956	$50,292,841
Texas	25,145,561	33,946,507
New York	19,378,102	26,160,438
Florida	18,801,310	25,381,769
Illinois	12,830,632	17,321,353
Pennsylvania	12,702,379	17,148,212
Ohio	11,536,504	15,574,280
Michigan	9,883,640	13,342,914
Georgia	9,687,653	13,078,332
North Carolina	9,535,483	12,872,902
New Jersey	8,791,894	11,869,057
Virginia	8,001,024	10,801,382
Washington	6,724,540	9,078,129
Massachusetts	6,547,629	$,839,299
Indiana	6,483,802	8,753,133
Arizona	6,392,017	8,629,223
Tennessee	6,346,105	8,567,242
Missouri	5,988,927	8,085,051
Maryland	5,773,552	7,794,295
Wisconsin	5,686,986	7,677,431
Minnesota	5,303,925	7,160,299
Colorado	5,029,196	6,789,415
Alabama	4,779,736	6,452,644
South Carolina	4,625,364	6,244,241
Louisiana	4,533,372	6,120,052
Kentucky	4,339,367	5,858,145

Table 15. (Continued)

Jurisdiction	2010 population	Total indicator amount (based on $1.35 per capita)a
Oregon	3,831,074	5,171,950
Oklahoma	3,751,351	5,064,324
Puerto Rico	3,725,789	5,029,815
Connecticut	3,574,097	4,825,031
Iowa	3,046,355	4,112,579
Mississippi	2,967,297	4,005,851
Arkansas	2,915,918	3,936,489
Kansas	2,853,118	3,851,709
Utah	2,763,885	3,731,245
Nevada	2,700,551	3,645,744
New Mexico	2,059,179	2,779,892
West Virginia	1,852,994	2,501,542
Nebraska	1,826,341	2,465,560
Idaho	1,567,582	2,116,236
Hawaii	1,360,301	1,836,406
Maine	1,328,361	1,793,287
New Hampshire	1,316,470	1,777,235
Rhode Island	1,052,567	1,420,965
Montana	989,415	1,335,710
Delaware	897,934	1,212,211
South Dakota	814,180	1,099,143
Alaska	710,231	958,812[a]
North Dakota	672,591	907,998[a]
Vermont	625,741	844,750[a]
District of Columbia	601,723	812,326[a]
Wyoming	563,626	760,895[a]
Guam	159,358	215,133[a]
Federated States of Micronesia	107,154	144,658[a,b]
Virgin Islands	106,405	143,647[a]
Marshall Islands	65,859	88,910[a,b]
American Samoa	55,519	74,951[a]
Northern Mariana Islands	53,883	$72,742[a]

Source: GAO analysis of U.S. Census and FEMA data.

[a]FEMA has established a minimum threshold of $1 million in PA damages per disaster in the belief that even the lowest-population states can cover this level of PA damage. 44 C.F.R. § 206.48(a)(1).

[b]The 2010 populations for the Federated States of Micronesia and the Marshall Islands are estimates from the U. S. Census Bureau.

APPENDIX IV: DESCRIPTION OF THREE APPROACHES TO MEASURE A JURISDICTION'S FISCAL CAPACITY

The Stafford Act requires that a governor's request for a major disaster declaration be based on a finding that the disaster is of such severity and magnitude that an effective response is beyond the capabilities of the jurisdiction and that federal assistance is necessary.[1] In the wake of a disaster, FEMA prepares dollar estimates of the damage to public infrastructure incurred in an area that would be eligible for federal assistance under a federal major disaster declaration. Currently the key metric for determining eligibility for federal disaster assistance is a "per capita indicator," which, since 1999, has been adjusted annually for inflation. For 2012, the indicator is $1.35 per capita (that is, total estimated damages eligible for federal PA divided by the jurisdiction's population equals $1.35 or more). Damage estimates in excess of this number typically result in FEMA recommending that the President issue a major disaster declaration, which makes jurisdictions eligible for federal reimbursement of at least 75 percent of certain repair and replacement costs. The per capita indicator FEMA currently uses is not a measure of a jurisdiction's fiscal capacity to address the damages caused by a disaster. Rather, there is an assumption that generally jurisdictions are unable without federal major disaster assistance to rectify damages that equal or exceed $1.35 per capita.

Jurisdictions' abilities to finance their own disaster relief and recovery vary with their fiscal capacity, among other factors. A jurisdiction's fiscal capacity is defined as the ability of a government to raise revenue from its own sources, by taxes, license fees, user charges, and public enterprises, among other devices. Fiscal capacity is usually expressed as a percentage of the national average for the 50 states plus the District of Columbia, in the form of an index number. For example, if a jurisdiction's capacity is equal to 100 percent or 90 percent of the national average, its index number would be 100 or 90, respectively. In general, the ability of jurisdictions to pay for public services increases with the size of their economies. The simplest application of fiscal capacity criteria to disaster assistance is an adjustment of the per capita indicator for every jurisdiction. "Richer" jurisdictions—those above the national average— would have a higher level, reflecting their greater ability to pay, while "poorer" jurisdictions would have a lower level. The fiscal capacity index could be converted into a percentage (for instance, 110 = 110 percent or 90 = 90 percent) and applied to each jurisdiction's per capita indicator ($1.35

times the population) to get an adjusted indicator. For example, a jurisdiction with a population of 10 million would have a per capita indicator of $13.5 million dollars ($1.35 times its 10 million population). If the jurisdiction's capacity index was 100 percent, its capacity threshold would be the same as the current indicator—$13.5 million. However, if the jurisdiction's fiscal capacity index was 110 (indicating a fiscal capacity index 10 percent above the national average), its damage threshold would be $14.85 million—$13.5 million plus 10 percent of $13.5 million. The variation in the results could be narrowed, if desired, by setting upper and lower limits to the adjustments, among other possible methods. However, adjusting the per capita indicator according to a jurisdiction's fiscal capacity would not necessarily reduce total federal spending. It is possible that disaster assistance adjusted by fiscal capacity, and focused on jurisdictions with below-average fiscal capacity, could increase total federal spending. How total annual spending is affected would depend on the specific disasters taking place for that year, as well as the affected jurisdictions.

In addition to the theoretical aspects of comparing and contrasting various measurements of fiscal capacity, there are other matters to consider.[2] Specifically, there are certain attributes that are desirable in a fiscal capacity measure. These attributes could help FEMA determine the extent to which the agency could use measurements of jurisdiction fiscal capacity when determining a jurisdiction's eligibility for federal assistance. These attributes include the following:

Simple and easy to calculate: For political acceptance and analytical ease, the methodology of measurement should be as simple as possible. For practicality and transparency, a measure should be easy and inexpensive to calculate.

Convenient, available, and timely: Ideally the data for a measure would be routinely collected, checked, and published by a government agency on a timely basis. For example, the measurement should be possible on an annual basis with as little a time lag as possible, in order to provide the most timely indicator of jurisdictions' capacities.

Comprehensive: A measure should be comprehensive with respect to the implementation of its approach. Incompleteness could bias results.

Analytically sound: The principal analytical choice in capacity measurement is between economic measures that aim for comprehensive measurement free of double-counting, and tax base measures that in some way take account of governments' choices in how to tax.[3] There is debate among economists as to which type of measurement (economic or tax base) is more analytically sound.

Does not affect or is not affected by any individual jurisdiction's fiscal choices: Capacity measures should not be affected by or affect a jurisdiction's actual fiscal choices, in terms of what to tax, how to tax, or how much to tax. In principle a government's fiscal behavior could affect its own tax bases. No capacity measure makes any adjustments based on the impact of taxes on a state's economy because of the extreme difficulty of arriving at a simple method of making such adjustment that would earn political consensus.

Fiscal Capacity Measurements

The three fiscal capacity measures discussed below provide various methods that can be used to determine a jurisdiction's fiscal capacity. Each of these measures has benefits and potential shortcomings regarding the extent to which they measure a jurisdiction's fiscal capacity.

State personal income (PCI): As a measurement of a jurisdiction's fiscal capacity, state personal income is simple, available, and timely. The personal income of all residents of a jurisdiction consists of labor earnings, proprietors' and partnership income, rent, interest, dividends, and transfers (public cash benefits). It is the most commonly used measure in the United States for federal grants. PCI is simple and familiar to most people, and it is routinely calculated and published by the federal government on a jurisdiction-by-jurisdiction basis. More local measures of personal income are less comprehensive, for lack of data. Some jurisdictions may choose to tax only part of income, or not to tax it at all.

PCI aims to be a comprehensive measure of residents' personal income, without regard to how they are taxed.

PCI is not a comprehensive measure of a jurisdiction's fiscal capacity and is affected by a jurisdiction's fiscal choices. The principal shortcoming of PCI is its failure to reflect a jurisdiction's ability to raise tax revenue from nonresidents, also known as tax exporting. For example, a jurisdiction government may tax nonresident commuters, property owners, shoppers, and tourists. The ability to export taxes varies sharply across jurisdictions. Also, data on one component of personal income, accrued capital gains, is not available. More generally, changes in asset values are not captured in any fiscal capacity measures, because of lack of data. Another missing element in PCI is the net income of a jurisdiction's government enterprises. An example is state-owned liquor stores, whose profits never pass through private hands. Another example is royalties paid to governments by extractive industries, such as oil, gas, and uranium. These scenarios amount to income received by a jurisdiction's residents, through their government. In one state (Alaska), a share of such revenue—a "bonus" payment—is paid directly to state residents, which would appear as part of PCI. To a limited extent, PCI is biased to the extent to which a jurisdiction government finances transfer payments with taxes on income, since this income is counted twice, once by source and the second time by receipt.

Gross state product (GSP): As a measure of a jurisdiction's fiscal capacity, GSP is simple, available, timely, and not affected by a jurisdiction's fiscal choices. Also called gross domestic product by state, GSP consists of all income "produced" in a state. It includes labor earnings of those who work in a jurisdiction, irrespective of their residence, the net income of business firms operating in a jurisdiction, indirect business taxes paid to the governments of the jurisdiction, and the output of the public sector (in national income accounting, government output is valued at cost). GSP partially captures the ability of a jurisdiction to export taxes, since it includes income received by nonresidents and by the residents' own governments directly. As with PCI, with the benefit of government publication, the data are available on a timely basis and are easily converted into a fiscal capacity index. It is less affected by a jurisdiction's fiscal choices than PCI because it does not double-count income.

Similar to PCI, GSP is not a comprehensive measure of a jurisdiction's fiscal capacity. More specifically, PCI includes income received by a jurisdiction's residents, but not income generated in a jurisdiction but received by nonresidents. GSP does not include income received by a jurisdiction's residents that originates elsewhere.

Total taxable resources (TTR): As a measure of a jurisdiction's fiscal capacity, TTR is comprehensive, available, and not affected by jurisdictions' fiscal choices. According to the Department of the Treasury (Treasury), the object of TTR is to capture the unduplicated sum of PCI and GSP that is susceptible to taxation by a jurisdiction's government.[4] By this means the entirety of income potentially exposed to taxation is measured. In practice the calculation is relatively simple. A jurisdiction's GSP is supplemented with income received by jurisdiction residents that originated in other jurisdictions. This would include the labor earnings of residents who commute to jobs in other jurisdictions, and the capital income (mainly interest, dividends, rent, and capital gains) of all jurisdiction residents due to asset holdings in other jurisdictions. It excludes indirect business taxes paid to the federal government (such as the payroll tax and federal excises). TTR is used in two grant programs— the Department of Health and Human Services' Substance Abuse and Mental Health Services Administration's block grant program and Community Mental Health Service—and is calculated by Treasury.[5]

TTR is a complex measure of a jurisdiction's fiscal capacity and is not as timely as other measures. A crude measure of TTR is obtained by simply averaging a jurisdiction's PCI and GSP. This was done for official TTR estimates between 1992 and 1997. Subsequently the calculations were based more closely on the original conceptual framework for TTR. At present there is a 2-year lag in the publication of TTR estimates by Treasury. As of May of 2012, the most recent year available is 2009. The primary reason for the delay in publishing TTR estimates is the need to wait for federal excise tax revenues and nontax liabilities, and federal civilian enterprise surpluses, to become available. These data are provided to Treasury in August or September for the year ending 20 months prior. According to the Chief of the Regional Product Branch, Regional Product Division, at the Bureau of Economic Analysis, it could be possible to speed up the availability of TTR by 2 or 3 months. In addition, TTR is less transparent than PCI or GSP. It relies on approximations of capital income (dividends, interest, and rent), since such quantities are not reported by place of origination. It does not discriminate among income flows according to the degree of susceptibility to taxation.

Examples of Fiscal Measures Applied to the Per Capita Indicator

Each of these three measures of a jurisdiction's fiscal capacity to respond to and recover from a disaster without federal assistance has its limitations and can affect each jurisdiction somewhat differently, compared with using the current $1.35 per capita damage estimate indicator. FEMA's current per capita indicator is simple and easy to understand, but it is not a measure of a jurisdiction's fiscal capacity. Nor does FEMA have a useful measure of a jurisdiction's response capabilities. All current measures of those capabilities are jurisdictions' self-reported data without reference to common metrics for assessing capability. Because FEMA's per capita indicator does not comprehensively assess a jurisdiction's response and recovery capabilities, including a jurisdiction's fiscal capacity, some combination of these measures could provide a more robust and useful assessment of a jurisdiction's capability to respond to and recover from a disaster without federal assistance, or with minimal federal assistance. This could include exploring the usefulness of supplementing the current damage indicator (which does not fully reflect changes in inflation since its adoption in 1986) with other measures of a jurisdiction's fiscal capacity and response capability.

For example, one potential alternative methodology could involve adjusting the per capita indicator for each jurisdiction based on a measure of jurisdiction fiscal capacity. If FEMA were to adjust the PA indicator for inflation, the adjusted PA indicator for fiscal year 2011 would be $2.07. Beginning with the adjusted PA indicator of $2.07, each jurisdiction's PA indicator could then be adjusted based on that jurisdiction's fiscal capacity. For example, if the $2.07 base were adjusted for "jurisdiction A," which has a 2009 TTR index of 71.8, jurisdiction A's PA indicator would be $1.49. If the $2.07 base were adjusted for "jurisdiction B's" 2009 TTR index of 149, jurisdiction B's PA indicator would be $3.08 (see table16). The variation in jurisdiction A's $1.49 indicator and jurisdiction B's $3.08 indicator represents the difference in the two jurisdiction's fiscal capacities in accordance with each jurisdiction's TTR. In making any changes or enhancements to the methods used to assess a jurisdiction's fiscal capacity, policymakers would need to consider the relative priority of key attributes, as previously discussed, and the benefits and costs of developing and implementing such changes.

Table 16. PA Indicator Adjusted by Jurisdiction for 2009 TTR

	PA indicator adjusted forinflation, as recommended	Jurisdictions' 2009 index for TTR	PA indicator as adjusted forjurisdiction's TTRfiscal capacity index
Jurisdiction A	$2.07	71.8	$1.49
Jurisdiction B	$2.07	149.0	$3.08

Source: GAO analysis of Department of Commerce, Department of Treasury, and Bureau of Labor Statistics data.

End Notes

[1] Hereafter in this report, major disaster declarations are referred to as disaster declarations. In addition to issuing major disaster declarations, the President may issue emergency declarations (42 U.S.C. § 5191) and fire assistance declarations (42 U.S.C. § 5187). If the President declares an emergency, the federal government may provide immediate and short-term assistance that is necessary to save lives, protect property and public health and safety,

or lessen or avert the threat of a catastrophe, among other things. 42 U.S.C. § 5192. Federal assistance may not exceed $5 million under an emergency declaration unless continued emergency assistance is immediately required; there is a continuing and immediate risk to lives, property, public health or safety; and necessary assistance will not otherwise be provided on a timely basis. 42 U.S.C. § 5193. Upon the request of a governor, the President may issue a fire assistance declaration that provides financial and other assistance to supplement state and local firefighting resources for fires that threaten destruction that might warrant a major disaster declaration. 44 C.F.R. § 204.21.

[2] Hereafter in this report, jurisdictions refer to U.S. states, the District of Columbia, U.S. territories, Freely Associated States (nations in free association with the United States under the Compacts of Free Association), and local governments. The IA program provides for the necessary expenses and serious needs of disaster victims that cannot be met through insurance or low-interest Small Business Administration loans. For example, FEMA may provide temporary housing assistance, counseling, unemployment compensation, or medical expenses incurred as a result of a disaster. The PA program provides for debris removal; emergency protective measures; and the repair, replacement, or restoration of disaster-damaged, publicly owned facilities and the facilities of certain private nonprofit organizations that provide services otherwise performed by a government agency.

[3] 44 C.F.R. § 206.47.

[4] No-year funds are available for obligation without fiscal year limitation. An obligation is a definite commitment that creates a legal liability of the government for the payment of goods and services ordered or received, or a legal duty on the part of the United States that could mature into a legal liability by virtue of actions on the part of the other party beyond the control of the United States. Payment may be made immediately or in the future. An agency incurs an obligation, for example, when it places an order, signs a contract, awards a grant, purchases a service, or takes other actions that require the government to make payments to the public or from one government account to another. GAO, *A Glossary of Terms Used in the Federal Budget Process*, GAO-05-734SP (Washington, D.C.: September 2005).

[5] Hazard Mitigation provides additional funds to states to assist communities in implementing long-term measures to help reduce the potential risk of future damages to facilities. Mission Assignments are costs paid by FEMA for work assigned to other agencies and departments.

[6] GAO, *Disaster Assistance: Improvement Needed in Disaster Declaration Criteria and Eligibility Assurance Procedures,* GAO-01-837 (Washington, D.C.: Aug. 31, 2001).

[7] GAO, *2012 Annual Report: Opportunities to Reduce Duplication, Overlap and Fragmentation, Achieve Savings, and Enhance Revenue*, GAO-12-342SP (Washington, D.C.: Feb. 28, 2012), 321–328.

[8] Since September 11, 2001, the federal government has provided billions of dollars to state and local governments for planning, equipment, and training to enhance the capabilities of first responders to respond to both smaller-scale natural disasters and terrorist attacks. However, the federal financial assistance provided in the last several years has not been guided by a clear risk-based strategic plan that outlines the role of federal, state, and local governments in identifying, enhancing, maintaining, and financing critical first responder capabilities for emergencies. See GAO, *21st Century Challenges: Reexamining the Base of the Federal Government*, GAO-05-325SP (Washington, D.C.: Feb. 1, 2005).

[9] In addition to conducting interviews with officials in FEMA headquarters, we conducted site visits in September 2011 to the two FEMA regions that had the highest total obligations during fiscal years 2004 through 2011. The regional administrative offices were located in

Atlanta, Georgia, and Denton, Texas, respectively. At each region, we interviewed the Regional or Deputy Administrator and various other regional personnel. In addition, we visited the emergency management agencies for Georgia and Oklahoma and interviewed various officials, including the directors for each of these agencies. We selected the Georgia and Oklahoma state emergency management agencies based on their respective proximity to FEMA's regional offices, their high level of experience with disasters, and their availability for a visit during September 2011. We wanted to avoid states that were actively responding to a disaster during that time. While the information we obtained on these site visits is not generalizable, the information provided important insights into the disaster declaration process.

[10] The PA per capita indicator is an amount of funding, $1.35 per capita for fiscal year 2012, that is multiplied by the population of the jurisdiction (for example, state) for which the governor is requesting a disaster declaration for PA, to arrive at a threshold amount, which is compared with the estimated amount of damage done to public structures. 44 C.F.R. § 206.48 (a)(1). FEMA uses the comparison as an indicator of the jurisdiction's need for federal assistance. For this analysis, we used 4 years of data (fiscal years 2008 through 2011) instead of 8 years of data (fiscal years 2004 through 2011) because PDA summaries are available in electronic format only back to fiscal year 2008. PDAs are conducted to identify the amount of damage a jurisdiction has incurred as a result of a disaster. Because of the associated workload and length of time required to provide PDA summaries in paper format, we did not request that FEMA provide PDA summaries prior to fiscal year 2008.

[11] GAO, *Standards for Internal Control in the Federal Government*, GAO/AIMD-00-21.3.1 (Washington, D.C.: November 1999).

[12] We examined fiscal years 1989 through 2011 because FEMA does not maintain detailed financial data on disaster declarations prior to fiscal year 1989.

[13] In 2008, we analyzed 83 disaster declarations to determine the reliability of FEMA's projections. We found that after the declarations had been open for 6 months, FEMA's projections were within roughly 10 percent of the amount that was ultimately obligated. While the results could not be generalized across all disaster declarations, the analysis provides valuable insight into the reliability of FEMA's projections after a declaration has been open for 6 months. As our projections data were as of April 30, 2012, all open disaster declarations from fiscal years 2004 through 2011 had been open for at least 6 months. Thus, we believe the projections are reliable for purposes of this report. Our 2008 analysis can be found within GAO, *Disaster Cost Estimates: FEMA Can Improve Its Learning from Past Experience and Management of Disaster-Related Resources,* GAO-08-301 (Washington, D.C.: February 2008).

[14] The administrative costs that we identify in this report were defined by FEMA as obligations from the DRF that support the delivery of disaster assistance.

[15] According to FEMA officials, administrative costs are typically higher as a percentage of assistance costs in the early months following a disaster, with the administrative cost percentage typically decreasing as the declaration matures. Thus, administrative costs for declarations with ongoing assistance, particularly recently declared disasters, could overstate final administrative costs as a percentage of total disaster assistance obligations. FEMA has three levels of status for disaster declarations—programmatically open, closed, and reconciled. Programmatically open means that all financial decisions are not completed and eligible work remains. Closed means that financial decisions have been made, but all projects are not complete. Reconciled means that all projects are complete and the FEMA-state agreement is closed.

[16] During this time, we reported preliminary observations on opportunities to reduce the costs to the federal government related to major disaster declarations. See GAO-12-342SP. 321-328.

[17] 42 U.S.C. § 5170.

[18] 42 U.S.C. § 5121 et seq.

[19] 6 U.S.C. § 317.

[20] "Emergency management" is defined as the managerial function charged with creating the framework within which communities reduce vulnerability to hazards and cope with disasters.

[21] Disaster Assistance Employees are a cadre of temporary reserve staff who go to field locations to help communities respond to and recover from disasters.

[22] The first presidential disaster declaration was issued in 1953 under the authority of the Disaster Relief Act of 1950. Pub. L. No. 81-875, 64 Stat. 1109.

[23] 42 U.S.C. § 5170. In addition to issuing major disaster declarations, the President may issue emergency declarations. If the President declares an emergency, the federal government may provide immediate and short-term assistance that is necessary to save lives, protect property and public health and safety, or lessen or avert the threat of a catastrophe, among other things. 42 U.S.C. § 5192. Federal assistance may not exceed $5 million under an emergency declaration unless continued emergency assistance is immediately required; there is a continuing and immediate risk to lives, property, public health or safety; and necessary assistance will not otherwise be provided on a timely basis. 42 U.S.C. § 5193.

[24] Federal agencies may provide assistance under Titles IV and V of the Stafford Act only upon a presidential declaration, but some federal agencies may respond to disasters under separate authorities.

[25] 44 C.F.R. § 206.36.

[26] The requirement for a joint PDA may be waived for those incidents of unusual severity and magnitude that do not require field damage assessments to determine the need for supplemental federal assistance or in other instances as determined by the Regional Administrator upon consultation with the state. 44 C.F.R. § 206.33.

[27] 44 C.F.R. § 206.48(b).

[28] In November 2009, FEMA implemented a new policy for disaster declarations for snowstorms. Prior to that, snowstorms were only declared as emergencies. The process for snowstorms is the same as that used for other types of disasters; however, a disaster declaration resulting from a snowstorm requires record or near-record snowfall—within 10 percent of a county's record snowfall as measured in inches. Covered costs for snowstorms include sanding, salting, de-icing, and removing snow.

[29] 44 C.F.R. § 206.46.

[30] The 5 territories are American Samoa, Guam, the U.S. Virgin Islands, the Commonwealth of the Northern Mariana Islands, and the Commonwealth of Puerto Rico. The 2 Freely Associated States are the Federated States of Micronesia and the Republic of the Marshall Islands.

[31] For the analysis of the jurisdictions with the lowest number of disaster declarations during fiscal years 2004 through 2011, we report on 4 jurisdictions because 6 jurisdictions tied for the fifth lowest, with 3 disaster declarations each.

[32] In addition to the $80.3 billion for the 539 disaster declarations during fiscal years 2004 through 2011, FEMA obligated funds for disaster declarations that occurred prior to fiscal year 2004 but were still receiving obligations during fiscal years 2004 through 2011. The $80.3 billion in obligations also excludes FEMA obligations for Emergency Declarations, Fire Management Assistance Grants, and non-DRF obligations of appropriated funds for

FEMA and federal agencies outside of DHS, such as the Department of Defense and Department of Health and Human Services. In addition to the 13 disasters that had exceeded $1 billion in obligations, other disasters declared during fiscal years 2004 through 2011 that were still open as of January 31, 2012, could reach obligations of over $1 billion as FEMA continues to obligate funds for them.

[33] We analyzed reconciled and closed disaster declarations because all financial decisions have been made for these declarations.

[34] Throughout this report, projected obligations refers to the actual amount obligated by FEMA as of April 30, 2012, plus the amount FEMA anticipates obligating between April 30, 2012, and the date a disaster declaration is ultimately reconciled. The information we analyzed included projections data for each declaration approved during fiscal years 1989 through 2011; however, the results within table 4 only focus on fiscal years 2004 to 2011.

[35] 76 Fed. Reg. 63,936 (Oct. 14, 2011).

[36] In addition to using a PA indicator for states, FEMA uses a PA indicator for counties to help determine a county's need for assistance. The countywide per capita indicator for fiscal year 2012 is $3.39. 76 Fed. Reg. 63,936 (Oct. 14, 2011).

[37] FEMA has established a minimum threshold of $1 million in PA damages per disaster in the belief that even the lowest population states can cover this level of public assistance damage. 44 C.F.R. § 206.48(a)(1).

[38] 42 U.S.C. § 5170. In this report, we consider a state's capabilities to respond to and recover from a disaster to include elements such as the fiscal capacity and preparedness of the state.

[39] 44 C.F.R. § 206.48.

[40] The Stafford Act prohibits a geographic area from being precluded from receiving assistance solely by virtue of an arithmetic formula or sliding scale based on income or population. 42 U.S.C. § 5163. We only reviewed FEMA's PDA summaries as far back as fiscal year 2008 because that is when FEMA began maintaining them in electronic format. For each of the 308 disaster declarations during fiscal years 2008 through 2011, we reviewed the PDA summary to determine whether a governor requested PA and the President approved PA, and whether the PA damage estimate met or exceeded the PA per capita indicator. Of the 308 PDA summaries, 246 of them included the data necessary to determine whether the PA damage estimate met or exceeded the PA per capita indicator.

[41] As discussed later in this report, the PA per capita indicator remained at $1.00 from 1986 through 1999 and has been adjusted for inflation on an annual basis since 2000.

[42] 51 Fed. Reg. 13,332 (Apr. 18, 1986).

[43] 64 Fed. Reg. 47,697 (Sept. 1, 1999).

[44] The Carrollton, Louisiana, water treatment facility was eligible for and did receive PA funding.

[45] FEMA determined that the National Cathedral was not eligible for PA; therefore the damage estimate was not included in the PDA and the cathedral was not eligible to receive PA funds. The epicenter of the earthquake was in Virginia.

[46] 42 U.S.C. § 5170.

[47] 42 U.S.C. § 5163.

[48] GAO/AIMD-00-21.3.1.

[49] Our analysis included FEMA's projected obligations as of April 30, 2012, for only those 508 disaster declarations that had received PA and had been declared during fiscal years 2004 through 2011. We did not analyze the 31 disaster declarations that received IA only. We analyzed obligations instead of PDA damage estimates for PA because FEMA officials stated that estimating the damage from a disaster is sometimes stopped when the estimate equals or exceeds the PA per capita indicator. Therefore, we concluded that conducting the

analysis using projected obligations would be more accurate than using incomplete PDA damage estimates for PA.

[50] In addition to our analysis of the 508 disaster declarations, we separately analyzed 144 of the 508 declarations wherein all financial decisions had been made as of January 31, 2012. Thus, we relied on actual obligations rather than projections to conduct the analysis. Our analysis of the 144 declarations showed that fewer disasters would have met either the personal income-adjusted or the inflation-adjusted PA per capita indicators for the years in which the disasters were declared. Specifically, our analysis showed that 49 percent of the 144 disaster declarations would not have met the PA per capita indicator if adjusted for the change in per capita personal income since 1986. Similarly, 31 percent of the 144 disaster declarations would not have met the PA per capita indicator if adjusted for inflation since 1986.

[51] In May 2012, the DHS Office of Inspector General reached a similar conclusion based on its analysis that, if FEMA had continually updated the indicator for changes in economic conditions, many recent disasters would not have met the financial statewide per capita indicator for federal assistance. See DHS Office of Inspector General, *Opportunities to Improve FEMA's Public Assistance Preliminary Damage Assessment Process*, OIG-12-709 (Washington, D.C.: May 2, 2012).

[52] 64 Fed. Reg. at 47,697 (Sept. 1, 1999); 64 Fed. Reg. 3910, 3911 (Jan. 26, 1999).

[53] GAO-01-837.

[54] 6 U.S.C. § 749.

[55] GAO, *Federal Emergency Management Agency: Continuing Challenges Impede Progress in Managing Preparedness Grants and Assessing National Capabilities*, GAO-12-526T (Washington, D.C.: Mar. 20, 2012).

[56] The Honorable W. Craig Fugate, FEMA Administrator, Testimony before the House of Representatives, Committee on Appropriations Subcommittee on Homeland Security, *Budget Hearing–Federal Emergency Management Agency–Director and State and Local Witnesses*, March 7, 2012.

[57] GAO-01-837.

[58] In addition to TTR, personal resident income, and gross state product, Representative Tax System is a measure of fiscal capacity. However, we did not include it as a potential means for FEMA to assess fiscal capacity because it is not currently calculated for U.S. jurisdictions, although it could be should FEMA choose to do so.

[59] 42 U.S.C. §§ 5170b(b), 5172(b), 5173(d). Generally, there is no nonfederal share for IA (FEMA's Other Needs Assistance program does require a cost share). 42 U.S.C. § 5174(g).

[60] 44 C.F.R. § 206.47. Instead of jurisdictions paying cash for their 25 percent or 10 percent share of eligible PA costs, they can use in-kind donations. According to FEMA, donated resources used on eligible work that is essential to meet immediate threats to life and property resulting from a major disaster may be credited toward the nonfederal cost share. For example, donated resources may include volunteer labor, equipment, materials, food, and shelter.

[61] Excluding the 23 legislatively required cost share adjustments from the 150 total requests means that FEMA's recommendation was a factor in 127 of the requested cost share adjustments. The President approved 86 of the 127 cost share adjustment requests, or 68 percent.

[62] See, e.g., Pub. L. No. 110-28, § 4501, 121 Stat. 112, 156 (2007) ("... the Federal share of assistance, including direct Federal assistance, provided for the States of Alabama, Florida, Louisiana, Mississippi, and Texas in connection with Hurricanes Katrina, Wilma, Dennis,

and Rita under sections 403, 406, 407, and 408 of the Robert T. Stafford Disaster Relief and Emergency Assistance Act... shall be 100 percent of the eligible costs under such sections.").

[63] 44 C.F.R. § 206.47(b).

[64] 44 CFR §206.47(d). Generally, 72 hours is the period of time that the federal government will pay 100 percent of these costs; however, the President can extend the time period.

[65] 44 C.F.R. § 206.48(b).

[66] GAO/RCED, *Disaster Assistance: Improvements Needed in Determining Eligibility for Public Assistance*, GAO-96-113 (Washington, D.C.: May 1996).

[67] GAO/AIMD-00-21.3.1.

[68] GAO/AIMD-00-21.3.1.

[69] In November 2010, FEMA established administrative cost percentage target ranges for disaster declarations. According to FEMA officials, the target ranges are not considered formal guidance and FEMA officials responsible for managing administrative costs are not held accountable for meeting the targets; however, the targets are supposed to shape how its leaders in the field think about gaining and sustaining efficiencies in operations. Event Level 1 declarations—with projected obligations of $500 million to $5 billion—have an administrative cost percentage target range of 8 percent to 12 percent of total obligations. Event Level 2 declarations—with projected total obligations from $50 million to $500 million—have a target range of 9 percent to 15 percent. Event Level 3 declarations— with projected obligations of less than $50 million—have a target range of 12 percent to 20 percent.

[70] According to FEMA officials, the agency relied on the Statement of Federal Financial Accounting Standards 4 to define its administrative costs. Agency officials calculate a declaration's administrative cost percentage by dividing administrative cost obligations by total obligations. Total obligations include funds obligated for PA, IA, Hazard Mitigation, Mission Assignment, and administration. We calculated administrative cost percentages using the formula provided by FEMA.

[71] In addition to FEMA's administrative costs, funds are obligated from the DRF to reimburse states and localities for certain administrative costs. Because these administrative costs are for state and local activities rather than FEMA activities, we did not include the costs in our administrative cost analyses. Further, certain administrative costs associated with disaster declarations are obligated from FEMA's annual appropriations. For example, according to FEMA officials, the salary of a full-time FEMA employee who works at headquarters is obligated from the agency's appropriations, even if the employee works on a specific disaster declaration. As FEMA could not quantify the amount of administrative costs obligated from its appropriations that related to disaster declarations, these costs were excluded from our administrative cost analyses.

[72] In addition to using JFOs, FEMA utilizes other temporary offices to respond to disaster declarations, including Disaster Recovery Centers and PA Processing Centers, which also incur administrative costs.

[73] FEMA uses three categories to characterize the status of a disaster declaration: programmatically open, closed, and reconciled. "Programmatically open" means that all financial decisions are not completed and eligible work remains. "Closed" means that financial decisions have been made, but all projects are not complete. "Reconciled" means that all projects are complete and the FEMA-state agreement is closed.

[74] Two of the 19 large disaster declarations had projected obligations over $5 billion, which is above the dollar range that FEMA uses to define large declarations. We included them in the group of large declarations.

[75] The administrative cost percentage is a percentage of the total obligation, not a percentage of the amount for disaster assistance. For example, if FEMA obligated $5 million for administrative costs and $5 million for disaster assistance, we divided the $5 million in administrative costs by $10 million ($5 million in administrative costs plus $5 million in disaster assistance). Therefore, in this example, the administrative cost percentage is 50 percent of the total $10 million obligation for that disaster.

[76] As discussed later in this report, FEMA offered some reasons why administrative cost percentages for IA are generally higher than for PA.

[77] FEMA, *Achieving Efficient JFO Operations: A Guide for Managing Staffing Levels and Administrative Costs* (Washington, D.C.: November 2010).

[78] We examined fiscal years 1989 through 2011 because FEMA maintains detailed financial data on disaster declarations only back to fiscal year 1989. To examine whether administrative costs changed over time, we divided the 23 years into three periods—two periods of 8 years each and one period of 7 years. FEMA officials stated that no significant changes have been made to the agency's definition of administrative costs during fiscal years 1989 through 2011.

[79] A virtual JFO is an off-site location wherein disaster workforce personnel utilize communication tools to deliver disaster assistance programs.

[80] The FCO is a position created by the Stafford Act and is appointed by the President to manage federal resource support activities related to disaster declarations. 42 U.S.C. § 5143. The FCO is responsible for coordinating the timely delivery of federal disaster assistance resources and programs to the affected localities.

[81] Statement of the Honorable W. Craig Fugate, Administrator, FEMA, before the House Committee on Transportation and Infrastructure, Subcommittee on Economic Development, Public Buildings, and Emergency Management, *Streamlining Emergency Management: Improving Preparedness, Response, and Cutting Costs,* U.S. House of Representatives (Washington, D.C.: October 13, 2011).

[82] Under the current reimbursement program, the state and local governments pay for the item or service and submit documentation to FEMA officials, who review the documentation to ensure it is an eligible expense and reimburse the state and local governments.

End Notes for Appendix I

[1] Figure 1 presented earlier in the report includes data on declarations since 1953, when the first presidential disaster declaration was issued.

End Notes for Appendix II

[1] The obligations data were current as of January 31, 2012. When finalizing our study, we received projected obligations data from FEMA that included updated actual obligations as of April 30, 2012. However, we did not receive detailed actual obligations data that would

have allowed us to update all of our actual obligations data analysis, so this analysis remained as of January 31, 2012.

[2] GAO, *Standards for Internal Control in the Federal Government*, GAO/AIMD-00-21.3.1 (Washington, D.C.: November 1999).

End Notes for Appendix IV

[1] 42 U.S.C. § 5170.

[2] Stephen M. Barro, "Macroeconomic Versus RTS Measures of Fiscal Capacity: Theoretical Foundations and Implications for Canada," Institute of Intergovernmental Relations, Working Paper, (Queens University, 2002), 1-5.

[3] The three fiscal measures we discuss in this appendix are economic measures. However, in addition to TTR, personal resident income, and gross state product, Representative Tax System (RTS) is a tax base measure. We did not include RTS as a potential means for FEMA to assess fiscal capacity because it is not currently calculated for U.S. jurisdictions, although it could be should FEMA choose to do so.

[4] U.S. Department of the Treasury, Office of Economic Policy, Treasury Methodology for Estimating Total Taxable Resources (TTR), (Washington, D.C.: revised November 2002). http://www.treasury.gov/resource-center/economic-policy/

[5] Michael L. Compson, "Historical Estimates of Total Taxable Resources for U.S. States, 1981-2000," *The Journal of Federalism,* 33:2 (Spring 2003), 55-72.

INDEX

F

G

H

I

J

L

M

N

O

S

T

U

V

W